This is the Statistics Doesn't W

So Easy, It's Practically Cheating...

4th edition

S. Deviant, MAT

StatisticsHowTo.com

States of America ISBN 978 1496163400

Contents

GLOSSARY

!	Factorial
f	Frequency
n	Number, or sample size
Q1	First Quartile
Q3	Third Quartile Variance
P(X)	Probability of an event, X
E	Margin of error
E(X)	Expected value
p	Probability of success (binomial distribution)
q	Probability of failure (binomial distribution)
μ	Mean, average
σ	Standard Deviation
s	Sample standard deviation
$\bar{X}$	Sample Mean
σ^2	Variance
α	Alpha level, significance level
H_0	Null Hypothesis
H_1	Alternate hypothesis
r	Correlation coefficient
r^2	Coefficient of determination
Σ	Sum of
a	Y intercept (linear regression)
b	Slope of the line (linear regression)
v	Degrees of freedom

Pre-Statistics

There are some elements from basic math and algebra classes that are must-knows in order to be successful in statistics. This section should serve as a gentle reminder of those topics.

HOW TO CONVERT FROM A PERCENTAGE TO A DECIMAL

"Percent" means how many out of one hundred. For example, 50% means 50 out of 100, 25% means 25 out of 100.

Step 1: *Move the decimal point two places to the left.*

Example: 75% becomes **.75%**.

Step 2: *Remove the "%" sign.*

.75% becomes **.75**.

HOW TO CONVERT FROM A DECIMAL TO A PERCENTAGE

Step 1: Move the decimal point two places to the right:

.75 becomes **75.00**
Step 2: Add a "%" sign.

75.00 becomes **75.00%**
Step 3 (Optional): You can drop the zeros if you want

75.00% becomes 75%

HOW TO ROUND TO X DECIMAL PLACES

Sample Problem: Round the number .1284 to 2 decimal places

Step 1: Identify where you want the number to stop by counting the number starting from the decimal point and moving right. One point to the right for the number .1284 would stop between the numbers 1 and 2. Two decimal places for the number .1284 will stop the number between 2 and 8:

.12|84.

Step 2: If the number to the right of that stop is 5 or above, round the number to the left up. In our example, .12|84, the 8 is larger than 5, so we round the left number up from 2 to 3:

.12|84 becomes **.13|84**.

Step 3: Remove any digits to the right of the stop:

.13|84 becomes **.13**.

Tip: Another way of looking at this is to take both numbers, either side of that stop mark I put in: 2|8 (we can forget the 4 at the end, as we're rounding way before that). Think of this as the number "28" and ask yourself: is this closer to 20, or closer to 30? It's closer to 30, so you can say .12|84, rounded to 2 decimal places, equals **.1300**. Then, you can just drop the zero.

HOW TO USE ORDER OF OPERATIONS

In order to succeed in statistics, you're going to have to be familiar with Order of Operations. The "operations": addition, multiplication, division, subtraction, exponents, and grouping, can be performed in multiple ways, and if you don't follow the "order", you're going to get the problem wrong. For example:

$$4 + 2 \times 3$$

...can equal either 18 or 10, depending on whether you do the multiplication first or last:

$4 + 2 \times 3$

6×3

18 (Wrong!)

$4 + 2 \times 3$

$4 + 6$

10 (Right!)

Here's how to get it right every time with an acronym: PEMDAS.

The acronym PEMDAS actually stands for Parentheses, Exponents, Multiplication, Division, Addition, Subtraction, but you may easily remember it as "Please Excuse My Dear Aunt Sally". Let's say we have a question that asks you to solve the following equation:

$$1.96 + z$$

Where:

$z = (\bar{X} - \mu) / (s / \sqrt{n})$

You are given:

$\bar{X} = 5$

$\mu = 3 \times 7$

$s = 5^2$

$n = 100$

Step 1: Insert the numbers into the equation:

$1.96 + z = 1.96 + ((5 - (3 \times 7)) / (5^2 / \sqrt{100}))*$
or
$1.96 + z = 1.96 + ((5 - (3 \times 7)) / 5^2 / 10))**$

Step 2: Parentheses. Work from the inside out if there are multiple parentheses. The inside parentheses is $3 \times 7 = 21$, so:

$1.96 + z = 1.96 + ((5 - 21) / (5^2 / 10))$

and there's only one operation left inside that first set of parentheses, so let's go ahead and do it:

$1.96 + z = 1.96 + ((-16) / (5^2 / 10))$

Next, we tackle the second set of parentheses:

Step 3: Exponents. $5^2 = 25$, so:

$1.96 + z = 1.96 + (-16 / (25 / 10))$

Step 4: There's no Multiplication in this equation, so let's do the next one: Division. Notice that there are two division signs. You may be wondering which one you should do first. Remember that we have to evaluate everything inside parentheses first, so we can't do the first division sign on the left until we evaluate (25 / 10), so let's do that:

$25 / 10 = 2.5$
$1.96 + z = 1.96 + (-16/(2.5))$

Now we can do the second division:

$-16 / 2.5 = -6.4$

$1.96 + z = 1.96 + (-6.4)$

Step 5: Addition.

1.96 + (-6.4) = -4.44

Step 6: Subtraction. In this case there is nothing more to do, so we're already done!

This should hopefully jog your memory—PEMDAS and Order of Operations is covered in basic math classes and you should be thoroughly familiar with the concept in order to succeed with statistics.

Notes:

* Notice that I placed parentheses around the top and bottom parts of the equation to separate them. If I hadn't, the equation would look like this:

$$z = 5 - 3 \times 7 / 52 / \sqrt{100}$$

which could easily be interpreted (incorrectly) as this:

$z = (5 - 3 \times 7 / 52) / \sqrt{100}$
So use parentheses liberally, as a reminder to yourself which parts of the equation go together.

** $\sqrt{100} = 10$. You don't need PEMDAS to evaluate simple square roots like this one.

HOW TO FIND THE MEAN, MODE AND MEDIAN

Step 1: *Put the numbers in order so that you can clearly see patterns.*
For example, let's say we have:

2, 19, 44, 44, 44, 51, 56, 78, 86, 99, 99

The **mode** is the number that appears the most often. In this case the mode is **44**—it appears three times.

Step 2: *Add the numbers up to get a total.* Example:

2 +19 + 44 + 44 +44 + 51 + 56 + 78 + 86 + 99 + 99 = 622

Set this number aside for a moment.

Step 3: *Count the amount of numbers in the series.*
In our example (2, 19, 44, 44, 44, 51, 56, 78, 86, 99, 99), we have 11 numbers.

Step 4: *Divide the number you found in Step 2 by the number you found in Step 3.* In our example:

$622 / 11 = 56.5\overline{4}$ *
is the **mean**, sometimes called the average.

If you had an odd number in Step 3, go to Step 5. If you had an even number, go to Step 6.

Step 5: *Find the number in the middle of the series.*
~~2,19,44,44,44,~~ **51,**~~56,78,86,99,99~~
In this case it's **51**—this is the **median**.

Step 6: *Find the middle two numbers of the series.* Example:

~~1, 2, 5, 6,~~ **7, 8,** ~~12, 15, 16, 17~~

The **median** is the number that comes in the middle of those two numbers (7 and 8). To find the middle, add the numbers together and divide by 2:

(7 + 8) / 2 = **7.5**

Tip: You can have more than one mode. For example, the mode of

1, 1, 5, 5, 6, 6 is **1, 5,** and **6**.

* An overline above decimal numbers means "repeating," so the number above is **56.54545454...** (going on forever).

WEIGHTED MEAN: FORMULA: HOW TO FIND WEIGHTED MEAN

A weighted mean is a kind of average. Instead of each data point contributing equally to the final mean, some data points contribute more "weight" than others. If all the weights are equal, then the weighted mean equals the arithmetic mean (the regular "average" you're used to). Weighted means are very common in statistics, especially when studying populations.

The Arithmetic Mean.

When you find a mean for a set of numbers, all the numbers carry an equal weight. For example, if you want to find the arithmetic mean of 1, 3, 5, 7, and 10:

Add up your data points: 1 + 3 + 5 + 7 + 10 = 26.

Divide by the number of items in the set: 26 / 5 = 5.2.

What do we mean by "equal weight"? You can think of each number in the sum above to contribute 1/5 to the total mean (as there are 5 numbers in the set).

The Weighted Mean.

In some cases, you might want a number to have more weight. In that case, you'll want to find the **weighted mean**. To find the weighted mean:

Multiply the numbers in your data set by the weights.

Add the numbers up.

For that set of number above with equal weights (1/5 for each

number), the math to find the weighted mean would be:
1(*1/5) + 3(*1/5) + 5(*1/5) + 7(*1/5) + 10(*1/5) = 5.2.

Sample Problem: You take three 100-point exams in your statistics class and score 80, 80 and 85. The last exam is *much* easier than the first two, so your professor has given it less weight. The weights for the three exams are:

Exam 1: 40 % of your grade. (Note: 40% as a decimal is .4.)

Exam 2: 40 % of your grade.

Exam 3: 20 % of your grade.

What is your final weighted average for the class?

Step 1: Multiply the numbers in your data set by the weights:
.4(80) = 32
.4(80) = 32

.2(85) = 19

Step 2: Add the numbers up. 32 + 32 + 19 = 83.

Weighted Mean Formula

The weighted mean is relatively easy to find. But in *some* cases the weights might not add up to 1. In those cases, you'll need to use the weighted mean formula. The only difference between the formula and the steps above is that you divide by the sum of all the weights.

$$\bar{x} = \dfrac{\sum\limits_{i=1}^{n}(x_i * w_i)}{\sum\limits_{i=1}^{n} w_i}$$

The image above is the technical formula for the weighted mean. In simple terms, the formula can be written as:

Weighted mean = Σwx/Σw

Σ = the sum of (in other words...add them up!).

w = the weights and x = the value.

To use the formula:

Step 1: Multiply the numbers in your data set by the weights.

Step 2: Add the numbers in Step 1 up. Set this number aside for a moment.

Step 3: Add up all of the weights.

Step 4: Divide the numbers you found in Step 2 by the number you found in Step 3.

In the sample grades problem above, all of the weights add up to 1 (.4 + .4 + .2) so you would divide your answer (83) by 1: 83 / 1 = 83.

However, let's say your weighted means added up to 1.2 instead of 1. You'd divide 83 by 1.2 to get: 83 / .8 = 69.17.

Warning: The weighted mean can be easily influenced by outliers in your data. If you have very high or very low values, the weighted mean may not be a good statistic to rely on.

HOW TO FIND A FACTORIAL (!)

Factorials are a concept usually introduced in elementary algebra. An exclamation mark indicates a factorial. Factorials are products of every whole number from 1 to the number before the exclamation mark. Examples:

$2! = 1 \times 2 = \mathbf{2}$

$3! = 1 \times 2 \times 3 = \mathbf{6}$

$4! = 4 \times 3 \times 2 \times 1 = \mathbf{24}$

Most calculators have a factorial (!) button, but Google can also do the work for you!

Step 1: *Go to the search bar at Google.com*

Step 2: *Type in a factorial, such as 36!* (that's an exclamation point)

Step 3: *Press enter*
Google can also figure out more complicated factorials for you, such as:

36! / (12 - 10)! 6!

However, make sure you put in parentheses and a multiplication sign (just as you would on any basic calculator):

$36! / ((12-10)! \times 6!) = \mathbf{2.58328699 \times 10^{38}}$

The Basics/Descriptive Statistics

HOW TO DETECT FAKE STATISTICS

How do you know whether to trust results from a survey or not? Do you believe an egg company when it tells you 90% of consumers in a taste test preferred their eggs? How about if a voluntary survey of U.S. marines showed overwhelming support for massive pay increases for military personnel? Sometimes it isn't enough to just accept the data as it is presented; dig a little deeper and you might uncover one of these common problems with statistics.

Step 1: Take a close look at **who paid** for the survey. If you read a statistic stating 90% of people lost 20 pounds in a month on a certain "miracle" diet, look at who paid for that survey. If it was the company who owns that "miracle" product, then it's likely you have what's called a **self-selection study**. In a self-selection study, someone stands to gain financially from the results of a trial or survey. You may have seen those soda ads where "90% of people prefer the taste of product X." But if the manufacturer of product X paid for that survey, you probably can't trust the results.

Step 2: Take a look at if the statistics came from a voluntary survey. A **voluntary response sample** is a sample where the participants can choose to be included in the sample or not. For example, if your professor was to send you an email with an invitation to comment on what you think of this book, then that would be a voluntary response sample. If it was a mandatory part of your course, then that would *not* be a voluntary response sample. Voluntary response samples are not suitable for statistics, because they carry a heavy bias toward people who have strong opinions (often negatives ones). In other words, students are more likely to respond to the above survey if they hate the textbook; the students who like the book will probably be less likely to respond.

Step 3: Look for the **faulty conclusion** that one variable causes another in the survey. For example, you might read a statistic that states unemployment causes an increase in corn production, because corn products (like high fructose corn syrup) are cheap, and therefore people are more likely to buy cheap foods when unemployed. But there may be many other factors causing an increase in production, including an increase in government subsidies for corn. Just because one factor is seemingly connected to another (correlation), that doesn't necessarily imply causation (that one caused the other)!

Step 4: Beware of **journal bias**. Journals are likely to report positive results (for example, a drug trial that had a positive outcome), rather than a drug trial that failed. Just because a journal publishes a positive result doesn't mean that there aren't other trials out there that reported a negative result.

Step 5: Make sure the **sample size** isn't too limited in scope. It's unlikely you can make generalizations about student achievement in the US by studying a single inner city school in Brooklyn. And it's unlikely you can make generalizations about American polling behavior by standing outside a polling booth in Ponte Vedra Beach, Florida. Just as inner city schools don't behave like every other school, an affluent neighborhood can't be used to generalize about the voting population. Also make sure the sample size is large enough: if your voting precinct contains 1 million voters, it's unlikely you'll get any good results from surveying 20 people.

Step 6: Watch out for **misleading percentages**. Unemployment may have "slowed by 50%," but if the unemployment rate was previously 100,000 new unemployment claims per month, that still means 50,000 people are joining the unemployed ranks every month.

Step 7: Beware of **precise numbers**. If a national survey reports that 3,150,023 households in the U.S. are dog owners, you might be inclined to believe that exact figure. However, it's highly unlikely (and almost impossible) that anyone would have seriously surveyed all of the households in the U.S. It's much more likely they surveyed a sample, and that 3,150,023 is an estimate (it should have been reported as 3 million to avoid being misleading). There are many other examples of fake statistics. Newspapers sometimes print erroneous figures, drug companies print fake test results, governments present skewed statistics in their favor. The golden rule is: **question every statistic that you read**.

HOW TO TELL THE DIFFERENCE BETWEEN DISCRETE AND CONTINUOUS VARIABLES

In an introductory stats class, one of the first things you'll learn is the difference between discrete and continuous variable statistics. How to tell the difference between the two:

Step 1: *Figure out how long it would take you to sit down and count out the possible values of your variable.* For example, if your variable is "Temperature in Arizona," how long would it take you to write every possible temperature? It would take you literally forever:

50°, 50.1°, 50.11°, 50.111°, 50.1111°, ...

If you start counting now and never, ever, ever finish (i.e. the numbers go on and on until infinity), you have what's called a **continuous variable**.

If your variable is "Number of Planets around a star," then you can count all of the numbers out (there can't be infinity planets). That is a **discrete variable**.

Step 2: *Think about "hidden" numbers that you haven't considered.* For example: is time discrete or continuous? You might think it's continuous (after all, time goes on forever, right?) but if we're thinking about numbers on a wristwatch (or a stop watch), those numbers are limited by the numbers or number of decimal places that a manufacturer has decided to put into the watch. It's unlikely that you'll be given an ambiguous question like this in your elementary stats class but it's worth thinking about!

How to Tell the Difference Between Different Sampling Methods

You'll come across many different types of samples in statistics: random samples, simple random samples, systematic samples, convenience samples, stratified samples, and cluster samples. Their names give away their meaning, so it's a straightforward process to determine what kind of sample you have.

Step 1: *Find out if the study sampled from individuals.* You'll find **random sampling** in a school lottery, where individual names are picked out of a hat randomly. A more "systematic" way of choosing people can be found in "**systematic sampling,**" where every Nth individual is chosen from a population. For example, every 100th customer at a certain store might receive a "doorbuster" gift.

Step 2: *Find out if the study picked groups of participants.* For large numbers of people (like the number of potential draftees in the Vietnam War), it's simpler to pick people by groups (**simple random sampling**). Draftees were grouped by birth date.

Step 3: *Determine if your study contained data from more than one carefully defined group* ("strata" or "cluster"). Some examples of strata could be: Democrats and Republics, Renters and Homeowners, or Country Folk vs. City Dwellers. If there are two very distinct, clear groups, you have a **stratified sample** or a "**cluster sample.**" If you have data about the individuals in the groups, that's a **stratified sample**. If you only have data about the groups themselves (you may only know the location of the individuals), then that's a **cluster sample**.

Step 4: *Find out if the sample was easy to get.* A classic example of **convenience sampling** is going to the mall and polling people who happen to walk by.

HOW TO TELL THE DIFFERENCE BETWEEN A STATISTIC AND A PARAMETER

A statistic and a parameter are very similar. They are both descriptions of groups, like "50% of dog owners prefer Tasty Brand dog food." The difference is that statistics describe a **sample**, whereas a parameter describes an entire **population**.

For example, if you randomly poll voters in a particular election and determine that 55% of the population plans to vote for candidate A, then you have a **statistic**: you only asked a sample of the population who they are voting for, then you calculated what the population was likely to do based on the sample.

Alternatively, if you ask a class of third graders who likes vanilla ice cream, and 90% of them raise their hands, then you have a parameter: 90% of that class likes vanilla ice cream. You know this because you asked *everyone* **in the population**.

Here are some steps you can take to be able to tell the difference between the two:

Step 1: *Ask yourself: is this obviously a fact about the whole population?* Sometimes that's easy to figure out. For example, with small populations, you usually have a parameter because the groups are small enough to measure:

- 10% of US senators voted for a particular measure. There are only 100 US Senators; you can count what every single one of them voted
- 40% of 1,211 students at a particular elementary school got below a 3 on a standardized test. You know this because you have each and every student's test score.
- 33% of 120 workers at a particular bike factory were paid less than $20,000 per year. You have the payroll data for all of the workers.

Step 2: *Ask yourself, is this obviously a fact about a very large population?* If it is, you have a statistic.

- 60% of US residents agree with the latest health care proposal. It's not possible to actually ask hundreds of millions of people whether they agree. Researchers have to just take samples and calculate the rest, so this is a statistic.
- 45% of Jacksonville, Florida residents report that they have been to at least one Jaguars game. It's very doubtful that anyone polled in excess of a million people for this data. They took a sample, so they have a statistic.
- 30% of dog owners poop scoop after their dog. It's impossible to survey all dog owners—no one keeps an accurate track of exactly how many people own dogs. This data has to be from a sample, so it's a statistic.

When in doubt, think about the time and cost involved in surveying an entire population. If you can't imagine anyone wanting to spend the time or the money to survey a large number (or impossible number) in a certain group, then you almost certainly are looking at a statistic.

HOW TO TELL THE DIFFERENCE BETWEEN QUANTITATIVE AND QUALITATIVE VARIABLES

In introductory statistics, it's easy to get confused when classifying a variable or object as quantitative or qualitative. Quantitative means it can be counted, like "number of people per square mile." Qualitative means it is a description, like "brown dog fur." How to classify those variables:

Step 1: *Think of a category for the items,* like "car models" or "types of potato" or "feather colors" or "numbers" or "number of widgets sold." The name of the category is not important.

Step 2: *Rank or order the items in your category.* Some examples of items that can be ordered are: number of computers sold in a month, students' GPAs or bank account balances. Anything with numbers or amounts can be ranked or ordered. If you find it impossible to rank or order your items, you have a qualitative item. Examples of qualitative items are "car models," "types of potato," "Shakespeare quotes."

Step 3: *Make sure you haven't added information.* For example, you could rank car models by popularity or expense, but popularity and expense are separate variables from "car model." If the item is "potatoes," it's qualitative. If the item is "number of potatoes sold," it's quantitative.

MAKE A HISTOGRAM IN EASY STEPS

Make a histogram: Overview

A histogram is a way of graphing groups of numbers according to how often they appear. This article will show you how to make one by hand, but you're *much* better off using technology, like making an Excel 2007 histogram. Choosing bins in statistics is usually a matter of an educated guess. When you make a histogram by hand, you're stuck with your original bin settings. With Excel, you can change the bins *after* you've created the histogram, giving you the ability to play around with bin sizes until you have a chart you're happy with. OK, enough of lecturing about technology. Sometimes you might *have* to make a histogram by hand, especially if you're making a relative frequency histogram; Technology like TI-83s will only create regular frequency histograms. If you *have* to create a histogram by hand, here's the easy way.

Make a Histogram: Steps

Sample question: Create a relative frequency histogram for the following test scores: 99, 97, 94, 88, 84, 81, 80, 77, 71, 25.

Step 1: Draw and label your x and y axis. For this example, the x-axis would be labeled "score" and the y-axis would be labeled "relative frequency %."

Step 2: Choose the number of bins (see: how to choose bin sizes in statistics) and label your graph. For this sample question, groups of 10 (the x-axis values are the bins) are a good choice (it looks like you'll have 5 bars of one or two items in the group).

Step 3: Divide 100 by the number of data points to get an idea of where to place your frequency "ticks". We have 10 items in our data set, so it makes sense to count by 100/10 = 10% (one item would equal 10% of the total).

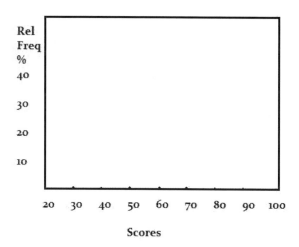

Step 4: Count how many items are in each bin and then sketch a rectangle on the graph that corresponds to the percentage of the total that bin fills. In this sample data set, the first bin (20-30) has 1 item; 70-80 has two items. If an item falls on a bin boundary (i.e. 80), place it in the *next* bin up (80 would go in 80-90).

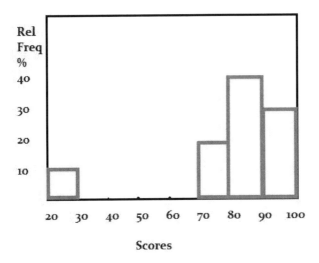

That's it!

Tip: Choosing where to place the frequency ticks is also somewhat of a judgment call and is rarely an exact science. For example, if you had 21 items, you could place your ticks at 5% although each item would be slightly less than 5%. Bear that in mind when you sketch the graph.

Warning: Choosing optimal bin sizes gets *very* complex with large data sets. The larger your data set, the better off you are using technology.

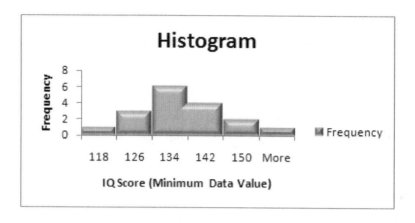

There isn't a formula to choose bin sizes in statistics. However, there are a few general rules:

Bins should be all the same size. For example, groups of ten or a hundred.

Bins should include *all* of the data, even outliers. If your outliers fall way outside of your other data, consider lumping them in with your first or last bin. This creates a rough histogram — make sure you note where outliers are being included.

Boundaries for bins should land at whole numbers whenever possible (this makes the chart easier to read).

Choose between 5 and 20 bins. The larger the data set, the more likely you'll want a large number of bins. For example, a set of 12 data pieces might warrant 5 bins but a set of 1000 numbers will probably be more useful with 20 bins. The exact number of bins is usually a judgment call.

If at all possible, try to make your data set evenly divisible by the number of bins. For example, if you have 10 pieces of data, work with 5 bins instead of 6 or 7.

Choose Bin size: Steps

Step 1: Find the smallest and largest data point. If your smallest and/or largest numbers are not whole numbers, go to Step 2. If they are whole numbers, go to Step 3.

Step 2: Lower the minimum a little and raise the maximum a little. For example, 1.2 as a minimum becomes 1, and 99.9 as a maximum becomes 100.

Step 3: Decide how many bins you need using your best guess and using the guidelines listed in the intro paragraph above.

Step 4: Divide your range (the number of items in your data set) by the bin size you chose in Step 3. For example, if you have 100 items and you chose 10 bins, your bin size is 10.

Step 5: Create the bin boundaries by starting with your smallest number (from Steps 1 and 2) and adding the bin size from Step 3. For example, if your smallest number is 1 and your bin size is 10 you would have bin boundaries of 1, 11, 21...

Tip: If you have a large data set, you may want to use Excel to find the smallest and largest point. Type your data into a single column and then use the "Sort" function or type =MIN(A:A) in a blank cell in a different column (i.e. column B) and then type =MAX(A:A) to get the biggest number.

RELATIVE FREQUENCY HISTOGRAM: DEFINITION AND HOW TO MAKE ONE

A relative frequency histogram is a type of graph that shows how often something happens, in percentages. The following relative frequency histogram shows book sales for a certain day. The price of the categories ("bins") are on the horizontal axis (the x-axis) and the relative frequencies (percentages of the whole) are shown in the vertical column (the y-axis).

Relative frequency histogram showing book sales for a certain day.

Relative frequency histograms are similar to pie charts, in that they show percentages of a whole. The total number of book sales in the above chart = .25 (25%) + .4 (40%) + .25 (25%) + .10 (10%) = 1 (100%).

How to Make a Relative Frequency Histogram: Steps

In these steps I'm going to show you by using a frequency chart to summarize your data. This is *optional* but it will really help you keep track of your data so that your histogram is correct!

Step 1: Make a frequency chart of your data. A frequency chart is just a list of the amount of times something happened. For the relative frequency histogram above, that would be the number of books sold for each price:

Price Range	Number Sold
$5-10	75
$11-15	120
$16-20	75
$21-$25	30

Step 2: Count the total number of items. For the sample problem, there were: 75 + 120 + 75 + 30 = 300 books sold.

Step 3: Figure out the relative frequency by dividing the count in each category by the total.
75/300 = .25
120/300 = .40
75/300 = .25
30/300 = .10
Add this information to a new column in your relative frequency chart.

Price Range	Number Sold	Relative .Frequency
$5-10	75	.25
$11-15	120	.40
$16-20	75	.25
$21-$25	30	.10

Step 4: Make a histogram with the information from your frequency chart. You *could* just sketch it by hand — putting the relative frequencies on the vertical axis and the prices on the horizontal axis (See: How to make a histogram). However, you can also use technology. I made the relative frequency histogram in this article with Excel.
That's it!

HOW TO MAKE A FREQUENCY CHART AND DETERMINE FREQUENCIES

If you are asked to determine a frequency in statistics, it doesn't just mean that you should just count out the number of times something occurs.

Step 1: *Make a chart for your data.* For this example, let's say you've been given a list of twenty blood types for incoming emergency surgery patients:

A, O, A, B, B, AB, B, B, O, A, O, O, O, AB, B, AB, AB, A, O, A

On the horizontal axis, write "frequency (#)" and "percent (%)". On the vertical axis, write your list of items. In this example, we have four distinct blood types: A, B, AB, and O.

	#	%
A		
B		
O		
AB		

Step 2: *Count the number of times each item appears in your data.* In this example, we have:

A appears 5 times

B appears 5 times

O appears 6 times

AB appears 4 times

Write those in the "number" column. This is your frequency.

	#	%
A	5	
B	5	
O	6	
AB	4	

Step 3: *Use the formula % = (f / n) × 100 to fill in the next column.* In this example, n = total amount of items in your data = **20**. A appears 5 times (frequency in this formula is just the number of times the item appears). So we have:

(5 / 20) × 100 = 25%

Fill in the rest of the frequency column, changing the 'f' for each blood type.

	#	%
A	5	25
B	5	25
O	6	30
AB	4	20

RIGHT SKEWED DISTRIBUTION: DEFINITION & EXAMPLES

A right skewed distribution is sometimes called a positively skewed distribution. That's because the tail is longer on the positive direction of the number line.

Right Skewed Histogram

A histogram is right skewed if the peak of the histogram veers to the left, giving the histogram's tail a positive skew to the right.

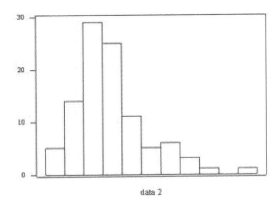

Image: SUNY Oswego

Right Skewed Box Plot

If a box plot is skewed to the right, the mean will be greater than the median. The box plot will look like the box was shifted to the left and the right whisker will be longer.

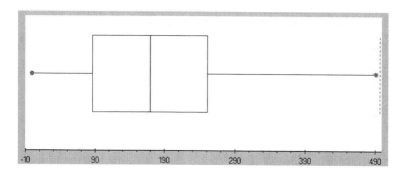

Image: Seton Hall University

Right Skewed Mean and Median

The rule of thumb is that in a right skewed distribution, the mean is usually to the right of the median.

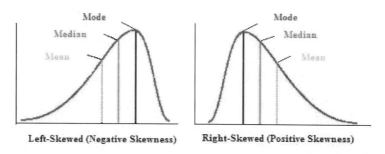

Left-Skewed (Negative Skewness)   Right-Skewed (Positive Skewness)

However, like most rules of thumb, there *are* exceptions. Most right skewed distributions you come across in elementary statistics will have the mean to the right of the median. The Journal of Statistics Education points out an exception to the rule:

In a data analysis course, the skew is often figured out using a third moment formula. Under these conditions, some distributions can break the rule of thumb. The following distribution was made from a 2002General Social Survey where respondents were asked how many people older than 18 live in their household. This is a right-skewed graph, but the mean is clearly to the left of the median.

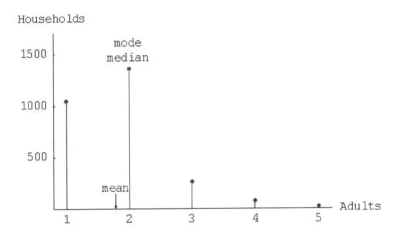

Image: Journal of Statistics Education

There are other exceptions which most involve theoretical mathematics and calculus. The important point to note is that although the mean is generally to the right of the median in a right skewed distribution, it isn't an absolute fact.

HOW TO FIND THE FIVE NUMBER SUMMARY BY READING A BOXPLOT

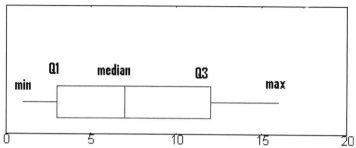

The five number summary consists of 5 items: The minimum, Q1 (the first quartile, or the 25% mark), The median, Q3 (the third quartile, or the 75% mark), The maximum.

The five-number summary gives you a rough idea about what your data set looks like. For example, you'll have your lowest value (the minimum) and the highest value (the maximum). Although it's useful in itself, the main reason you'll want to find a five-number summary is to find more useful statistics, like the interquartile range, sometimes called the middle fifty.

Step 1: *Find the Minimum.* The minimum is the far left hand side of the graph, at the tip of the left whisker. For this graph, the left whisker end is at approximately **.75**.

Step 2: *Find Q1.* Q1 is represented by the far left hand side of the box. In this case, about **2.5**.

Step 3: *Find the median.* The median is represented by the vertical bar. In this boxplot, it can be found at about **6.5**.

Step 4: *Find Q3.* Q3 is the far right hand edge of the box, at about **12** in this graph.

Step 5: *Find the maximum.* The maximum is the end of the "whiskers." In this graph, it's approximately **16**.

HOW TO FIND A FIVE NUMBER SUMMARY BY HAND

Step 1: *Put your numbers in order.* Example:

1, 2, 5, 6, 7, 9, 12, 15, 18, 19, 27

Step 2: *Find the minimum and maximum.* These are the first and last numbers in your ordered list. From the above example, the minimum is **1** and the maximum is **27**.

Step 3: *Find the median.* The median is the middle number of the list. In this case we have 11 total numbers, so the middle one is the sixth, which is **9**.

Step 4: *Place parentheses around the numbers above and below the median.* Not technically necessary, but it makes Q1 and Q3 easier to find.

(1, 2, 5, 6, 7), 9, (12, 15, 18, 19, 27)

Step 5: *Find Q1 and Q3.* Q1 can be thought of as a median in the lower half of the list, and Q3 can be thought of as a median for the upper half of data.

(1, 2, **5**, 6, 7), **9**, (12, 15, **18**, 19, 27)

In this case, Q1 is **5**, and Q3 is **18**.

HOW TO FIND AN INTERQUARTILE RANGE

Imagine all the data in a set as points on a number line. For example, if you have 3, 7 and 28 in your set of data, imagine them as points on a number line that is centered on 0 but stretches both infinitely below zero and infinitely above zero. Once plotted on that number line, the smallest data point and the biggest data point in the set of data create the boundaries of an interval of space on the number line that contains all data points in the set. The interquartile range (IQR) is the length of the middle 50% of that interval of space.

Sample Problem: Find the Interquartile range of the numbers: 9, 5, 12, 19, 2, 18, 7, 15, 1, 27, 6.

Step 1: *Put the numbers in order:*
1, 2, 5, 6, 7, 9, 12, 15, 18, 19, 27

Step 2: *Find the median:*
1, 2, 5, 6, 7, **9**, 12, 15, 18, 19, 27
The median is **9**.

Step 3: *Place parentheses around the numbers above and below the median.* Not necessary statistically, but it makes Q1 and Q3 easier to spot.
(1, 2, 5, 6, 7), 9, (12, 15, 18, 19, 27)

Step 4: *Find Q1 and Q3.* Q1 can be thought of as a median in the lower half of the list, and Q3 can be thought of as a median for the upper half of list.
(1, 2, **5**, 6, 7), **9**, (12, 15, **18**, 19, 27) Q1 is **5**, and Q3 is **18**.

Step 5: *Subtract Q3 from Q1 to find the interquartile range.*
18 − 5 = **13**

HOW TO FIND AN INTERQUARTILE RANGE ON A BOXPLOT

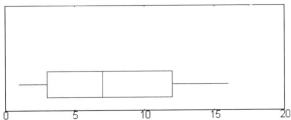

Step 1: *Find Q1.* Q1 is represented on a boxplot by the left hand edge of the "box." In the graph below, Q1 is approximately **2.5**.

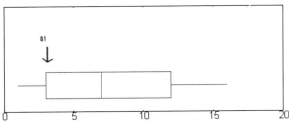

Step 2: *Find Q3.* Q3 is represented on a boxplot by the right hand edge of the "box." Q3 is approximately **12**.

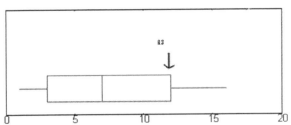

Step 3. *Subtract the number you found in Step 1 from the number you found in Step 3.* This will give you the interquartile range.

12 - 2.5 = **9.5**

HOW TO FIND THE SAMPLE VARIANCE AND STANDARD DEVIATION

The variance and standard deviations of samples are ways to measure the spread—or variability—of a sample or how far away from the "norm" a result is. In order to determine standard deviation, you have to find the variance first; standard deviation is just the square root of variance. The variance formula is:

$$\sigma^2 = \frac{\Sigma(x - \mu)^2}{N}$$

You could simply plug in the numbers. However, it can be tricky to use—especially if you are rusty on order of operations. By far the easiest way to find the variance and standard deviation is to use an online calculator (see the Resources chapter for links). However, if you don't have internet access when you are working these problems, here is how to find the variance and standard deviation of a sample in a few short Steps.

Step 1: *Add up the numbers in your given data set.* For example, let's say you were given a set of data for trees in California (heights in feet):

3, 21, 98, 203, 17, 9

Add them together:

3 + 21 + 98 + 203 + 17 + 9 = **351**

Step 2: *Square your answer:*

351 × 351 = **123,201**

...and divide by the number of items in your data set. We have 6 items in our example so:

123,201 / 6 = 20,533.5

Set this number aside for a moment.

Step 3: *Take your set of original numbers from Step 1, and square them individually this time:*

3 × 3 + 21 × 21 + 98 × 98 + 203 × 203 + 17 × 17 + 9 × 9

Add those numbers (the squares) together:

9 + 441 + 9604 + 41209 + 289 + 81 = **51,633**

Step 4: *Subtract the amount in Step 2 from the amount in Step 3.*

51,633 − 20,533.5 = **31,099.5**

Set this number aside for a moment.

Step 5: *Subtract 1 from the number of items in your data set.* For our example:

6 − 1 = **5**

Step 6: *Divide the number in Step 4 by the number in* **Step 5:**

31,099.5 / 5 = **6,219.9**

Step 7: *Take the square root of your answer from Step 6.* This gives you the standard deviation:

√6,219.9 = **78.86634**

HOW TO FIND A COEFFICIENT OF VARIATION

Sometimes we want to compare variations from different sets of data. Comparing variations in data isn't a problem if you are comparing two sets of IQ scores from similar classes, or two sets of SAT scores from incoming freshmen. But if you want to compare two sets of data that have different units (like two tests on different scales), then you need the coefficient of variation (CV). It's used to compare the standard deviations of two sets of data that have significantly different means.

Use the following formula to calculate the CV by hand for a population or a sample.

CV for a population: $CV = \frac{\sigma}{\mu} * 100\%$

CV for a sample: $CV = \frac{s}{\bar{X}} * 100\%$

σ is the standard deviation for a population, which is the same as "s" for the sample.
μ is the mean for the population, which is the same as $\bar{X}$ in the sample.

In other words, to find the coefficient of variation, divide the standard deviation by the mean and multiply by 100%.

How to find a coefficient of variation in Excel.

You can calculate the coefficient of variation in Excel using the formulas for standard deviation and mean. For a column of data (i.e. A1:A10), you could enter: =stdev(A1:A10)/=average(A1:A10).

How to Find a Coefficient of Variation by hand: Steps.

Sample question: Two versions of a test are given to students. One test has pre-set answers and a second test has randomized answers. Find the coefficient of variation.

	Regular Test	Randomized Answers
Mean	50.1	45.8
SD	11.2	12.9

Step 1: Divide the standard deviation by the mean for the first sample: 11.2 / 50.1 = 0.22355

Step 2: Multiply Step 1 by 100:
0.22355 * 100 = 22.355%

Step 3: Divide the standard deviation by the mean for the second sample:
12.9 / 45.8 = 0.28166

Step 4: Multiply Step 3 by 100:
0.28166 * 100 = 28.266%

That's it! Now you can compare the two results directly.

HOW TO DRAW A FREQUENCY DISTRIBUTION TABLE

Sample Problem: The IQ scores of the children in a particular gifted classroom were recorded as 118, 123, 124, 125, 127, 128, 129, 130, 130, 133, 136, 138, 141, 142, 149, 150, 154. Draw a frequency distribution table of these data.

Step 1: Figure out how many classes (categories) you need. There are no hard rules about how many classes to pick, but there are a couple of general guidelines:

Pick between 5 and 20 classes. For the list of IQs above, we picked 5 classes.

Make sure you have a few items in each category. For example, if you have 20 items, choose 5 classes (4 items per category), not 20 classes (which would give you only 1 item per category).

Step 2: Subtract the minimum data value from the maximum data value. For example, our the IQ list above had a minimum value of 118 and a maximum value of 154, so:
154 – 118 = **36**

Step 3: Divide your answer in Step 2 by the number of classes you chose in Step 1.
36 / 5 = **7.2**

Step 4: Round the number from Step 3 up to a whole number to get the class width. Rounded up, 7.2 becomes **8**.

Step 5: Write down your lowest value for your first minimum data value:
The lowest value is **118**

Step 6: Add the class width from Step 4 to Step 5 to get the next lower class limit:

118 + 8 = **126**

Step 7: Repeat Step 6 for the other minimum data values (in other words, keep on adding your class width to your minimum data values) until you have created the number of classes you chose in Step 1. We chose 5 classes, so our 5 minimum data values are:
118
126 (118 + 8)
134 (126 + 8)
142 (134 + 8)
150 (142 + 8)

Step 8: Write down the upper class limits. These are the highest values that can be in the category, so in most cases you can subtract 1 from the class width and add that to the minimum data value. For example:
118 + (8 − 1) = 125
118 − 125
126 − 133
134 − 142
143 − 149
150 − 157

Step 9: Add a second column for the number of items in each class, and label the columns with appropriate headings:

IQ	Number
118 − 125	
126 − 133	
134 − 142	
143 − 149	
150 − 157	

Step 10: Count the number of items in each class, and put the total in the second column. The list of IQ scores are: 118, 123, 124, 125, 127, 128, 129, 130, 130, 133, 136, 138, 141, 142, 149, 150, 154.

IQ	Number
118 – 125	4
126 – 133	6
134 – 142	4
143 – 149	1
150 – 157	2

Tip: If you are working with large numbers (like hundreds or thousands), round Step 4 up to a large whole number that's easy to make into classes, like 100, 1000, or 10,000.

HOW TO DRAW A CUMULATIVE FREQUENCY DISTRIBUTION TABLE

In elementary statistics you might be given a histogram and asked to determine the cumulative frequency. Or, you might be given a frequency distribution table and asked to find the cumulative frequency. The method for both is the same.

Sample Problem: Build a cumulative frequency table for the following classes.

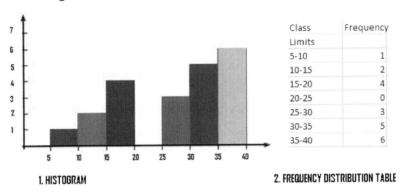

Class Limits	Frequency
5-10	1
10-15	2
15-20	4
20-25	0
25-30	3
30-35	5
35-40	6

1. HISTOGRAM   2. FREQUENCY DISTRIBUTION TABLE

Step 1: *If you have a histogram, go to Step 2. If you have a frequency distribution table (or both), go to Step 3.*

Step 2: *Build a frequency distribution table, like the one to the right of the histogram above.* Label column 1 with your class limits. In column 2, count the number of items in each class and fill the columns in as shown above. To fill in the columns, count how many items are in each class, using the histogram.

Step 3: *Label a new column in your frequency distribution table "Cumulative frequency" and compute the first two entries.* The first entry will be the same as the first entry in the frequency column. The second entry will be the sum of the first two entries in the frequency column.

Class	Frequency	Cumulative
5-10	1	1
10-15	2	3
15-20	4	
20-25	0	
25-30	3	
30-35	5	
35-40	6	

Step 4: *Fill in the rest of the cumulative frequency column.* The third entry will be the sum of the first three entries in the frequency column, the fourth will be the sum of the first four entries in the frequency column etc.

Class	Frequency	Cumulative
5-10	1	1
10-15	2	3
15-20	4	7
20-25	0	7
25-30	3	10
30-35	5	15
35-40	6	21

Probability

HOW TO CONSTRUCT A PROBABILITY DISTRIBUTION

Sample Problem: Construct a probability distribution for the following scenario: The probability of a sausage making machine producing 0, 1, 2, 3, 4, or 5 misshapen sausages per day are .09, .07, .1, .04, .12, and .02.

Step 1: *Write down the number of "widgets" given on one horizontal line.* In this case, we have "misshapen sausages."

Step 2: *Directly underneath the first line, write the result of the given function.* For example, the probability of the sausage machine producing 0 misshapen sausages a day is .09, so write ".09" directly under "0."

Number of misshapen sausages X	0	1	2	3	4	5
Probability P(X)	.09	.07	.1	.04	.12	.02

Fundamental Counting Principle: How to use it
Fundamental Counting Principle Definition.

The Fundamental Counting Principle is a way to figure out the number of outcomes in a probability problem. Basically, you multiply the events together to get the total number of outcomes. The formula is:

If you have an event "a" and another event "b" then all the different outcomes for the events is a * b.

Fundamental counting principle: Sample problem #1

A fast-food restaurant has a meal special: $5 for a drink, sandwich, side item and dessert. The choices are:

Sandwich: Grilled chicken, All Beef Patty, Vegeburger and Fish Filet.

Side: Regular fries, Cheese Fries, Potato Wedges.

Dessert: Chocolate Chip Cookie or Apple Pie.

Drink: Fanta, Dr. Pepper, Coke, Diet Coke and Sprite.

Q. How many meal combos are possible?
A. There are 4 stages:

Choose a sandwich.

Choose a side.

Choose a dessert.

Choose a drink.

There are 4 different types of sandwich, 3 different types of side, 2 different types of desserts and five different types of drink.

The number of meal combos possible is 4 * 3 * 2 * 5 = 120.

Fundamental counting principle: Sample problem #2.

Q. You take a survey with five "yes" or "no" answers. How many different ways could you complete the survey?

A. There are 5 stages: Question 1, question 2, question 3, question 4, and question 5.
There are 2 choices for each question (Yes or No).
So the total number of possible ways to answer is:
2 * 2 * 2 * 2 * 2 = 32.

Fundamental counting principle: Sample problem #3.

Q: A company puts a code on each different product they sell. The code is made up of 3 numbers and 2 letters. How many different codes are possible?
A. There are 5 stages (number 1, number 2, number 3, letter 1 and letter 2).
There are 10 possible numbers: 0 – 9.
There are 26 possible letters: A – Z.
So we have:
10 * 10 * 10 * 26 * 26 = 676000 possible codes.

HOW TO FIND THE PROBABILITY OF SELECTING A PERSON FROM A GROUP OR COMMITTEE

Problems about picking people from groups or committees are as ubiquitous in probability and statistics as pajamas at a sleepover. It doesn't matter whether your problem has doctors and nurses, committees or boards, parents and teachers or some other group of people: the steps are all the same!

Sample Problem: At a school board meeting there are 9 parents and 5 teachers. 2 teachers and 5 parents are female. If a person at the school board meeting is selected at random, find the probability that the person is a parent or a female.

Step 1: *Make a chart of the information given.* In the Sample problem, we're told that we have 5 female parents, 2 female teachers, 9 total parents and 5 total teachers.

Member	Female	Male	Total
Parent	5		9
Teacher	2		5

Step 2: *Fill in the blank column(s).* For example, we know that if we have 9 total parents and 5 are female, then 4 must be male.

Member	Female	Male	Total
Parent	5	4	9
Teacher	2	3	5

Step 3: *Add a second total to your chart* to add up the columns.

Member	Female	Male	Total
Parent	5	4	9
Teacher	2	3	5
Total	7	7	14

Step 4: *Add up the probabilities.* In our case we were asked to find out the probability of the person being a female or a parent. We can see from our chart that the probability of being a parent is 9/14 and the probability of being a female is 7/14:

9/14 + 7/14 = **16/14**

Step 5: *Subtract the probability of finding both at the same time.* If we don't subtract this, then female parents will be counted twice, once as females and another as parents.

16/14 – 5/14 = **11/14**

HOW TO FIND THE PROBABILITY OF AN EVENT **NOT** HAPPENING

Probability questions are often word problems that appear more difficult than they actually are. The trick is to identify the type of probability problem you have.

Step 1: *Choose one of the following Sample problems below that most closely matches your problem and go to the Step indicated.* Pay particular attention to the **bold** words when selecting.

"In a certain year, 12,942 construction workers were injured by falling debris. The top six cities for injuries were: New York (141), Houston (219), Miami (112), Denver (140), and San Francisco (110). If an injured construction worker is **selected at random**, what is the probability that the selected worker will have been injured in a city **other than** Houston, Miami, or Denver?" *If your problem looks like this, go to Step 2.*

"If the **probability** that a major hurricane will hit Florida this year is .49, what is the probability of a major hurricane **not** hitting Florida this year?" *If your problem looks like this, go to Step 5.*

"Twenty seven **percent** of American adults support Universal Health Care. If an American adult is chosen at random, what is the probability they **will not** support Universal Health Care?" *If your problem looks like this, go to Step 6.*

Step 2: *Add up the given cities.*

Houston (219) + Miami (112) + Denver (140) = **471**

Step 3: *Divide the number you found in Step 2 by the total number of people or items given.* In this case, we were told that 12,942 construction workers were injured:

471 / 12,942 = **.0364**

Step 4: *Subtract the number you found in Step 3 from the number*

1:

1 - .0364 = **.9936**

You're done!

Step 5: *Subtract the "not" probability from 1*:

1 - .49 = **.51**

You're done!

Step 6: *Convert the percentage to a decimal:*

Twenty seven percent = **.27**

Step 7: *Subtract the number in Step 6 from the number 1:*

1 - .27 = **.73**

You're done!

HOW TO SOLVE A QUESTION ABOUT PROBABILITY USING A FREQUENCY DISTRIBUTION TABLE

Sample Problem: In a sample of 43 students, 15 had brown hair, 10 had black hair, 16 had blond hair, and 2 had red hair. Use a frequency distribution table to find the probability a person has neither red nor blond hair.

Step 1: *Make a table.* List the items in one column and the tally in a second column.

Type	Frequency
Brown	15
Black	10
Blond	16
Red	2

Step 2: *Add up the totals.* In the Sample problem we're asked for the odds a person will **not** have blond or red hair. In other words, we want to know the probability of a person having black or brown hair.

Brown = 15 of 43

Black = 10 of 43

15/43 + 10/43 = **25/43**

HOW TO FIND THE PROBABILITY OF A SIMPLE EVENT HAPPENING

Probability questions are word problems, and they can all be broken down into a few simple steps.

Sample Problem: 50% of the families in the US had no children living at home. 22% had one child. 22% had two children. 4% had three children. 2% had four or more children. If a family is selected at random, what is the probability that the family will have three or more children?

Step 1: *Identify the individual probabilities and change them to decimals.* 4% of families have two children, and 2% have four or more children. Our individual probabilities are **.04** and **.02**.

Step 2: *Add the probabilities together.*

$$.04 + .02 = .06$$

Easy!

FINDING THE PROBABILITY OF A RANDOM EVENT GIVEN A PERCENTAGE

Here's how to solve a problem that gives a percentage (i.e. 76% of Americans are in favor of Universal Health Care), then asks you to calculate the probability of picking a certain number (i.e. 3 people) and having them *all* fall into a particular group (in our case, they are in favor of health care).

Step 1: *Change the given percentage to a decimal.* In our example:

76% = **.76**

Step 2: *Multiply the decimal found in Step 1 by itself.* Repeat for as many times as you are asked to choose an item, in our example, 3:

.76 × .76 × .76 = **.438976**

HOW TO FIND THE PROBABILITY OF GROUP MEMBERS CHOOSING THE SAME THING

These probability questions give you a group, and ask you to calculate the probability of an event occurring for all of a certain number of random members within that group.

Sample Problem: *There are 200 people at a book fair; 159 out of them will buy at least one book. If you survey 5 random people coming out of the door, what is the probability they all will have purchased one book?*

Step 1: *Convert the data in the question to a fraction.* For example, the phrase "159 people out of 200" can be converted to **159/200**.

Step 2: *Multiply the fraction by itself. Repeat for however many random items (i.e. people) are chosen.* In this example, we have 5 people surveyed:

$159/200 \times 159/200 \times 159/200 \times 159/200 \times 159/200 = .3176$

HOW TO FIND THE PROBABILITY OF TWO EVENTS OCCURRING TOGETHER

Sample Problem: 85% of employees have health insurance; out of this group, 45% had deductibles higher than $1000. How many people had deductibles higher than $1,000?

The key to these problems is the phrase "Out of this group" or "Of this group" which tells you that the events are dependent. You can solve these probability questions in only two Steps.

Step 1: *Convert your percentages of the two events to decimals.* In the above example:

$$85\% = .85$$

$$45\% = .45$$

Step 2: *Multiply the decimals from Step 1 together:*

$$.85 \times .45 = .3825$$

HOW TO FIND THE PROBABILITY OF AN EVENT, GIVEN ANOTHER EVENT

What these problems are asking you is 'given a certain situation, what is the probability that something else will occur?' These are called **dependent events**.

Sample Problem: Using the table of survey results below, find the probability that the answer was no, given that the respondent was male.

Gender	Yes	No	Total
Male	15	25	40
Female	10	50	60
Total	25	75	100

Step 1: *Find the probability of both the events in the question happening together.* In our sample, question, we were asked for the probability of the answer no from a male. From the table, the number of males who answered no is 25. More steps are needed to find the probability:

Step 2: *Divide your answer in Step 1 by the total figure.* In our example, it's a survey, so we need the total number of respondents (100, from the table):

25 / 100 = **.25**

Step 3: *Identify which event happened first (i.e. find the independent variable).* In our example, we identified the male subgroup and then we deduced how many answered no, so "total number of males" is the independent event. The question usually gives this information away by telling you "**given that** the respondent was male..." (as in our question).

Step 4: *Find the probability of the event in Step 3.* In our example, we want the probability of being a male in the survey. There are 40 males in our survey, and 100 people total, so the probability of being a male in the survey is:

40 / 100 = **.4**

Step 5: *Divide the figure you found in Step 2 by the figure you found in **Step 4:***

.25 / .4 = **.625**

HOW TO USE A PROBABILITY TREE FOR PROBABILITY QUESTIONS

Drawing a probability tree (or a tree diagram) is a way for you to visualize all of the possible choices, and to avoid making mathematical errors.

Sample Problem: An airplane manufacturer has three factories A, B, and C which produce 50%, 25%, and 25%, respectively, of a particular airplane. Seventy percent of the airplanes produced in factory A are passenger airplanes, 25% of those produced in factory B are passenger airplanes, and 25% of the airplanes produced in factory C are passenger airplanes. If an airplane produced by the manufacturer is selected at random, calculate the probability the airplane will be a passenger plane.

Step 1: *Draw and label lines to represent the first set of options in the question.* In our example, the options are the 3 factories, A, B, and C.

Step 2: *Convert the percentages to decimals,* and place those on the appropriate branch in the diagram. For our example:

50% = .5, 25% = .25

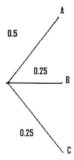

Step 3: *Draw the next set of branches.* In our case, we were told that 70% of factory A's output was passenger. Converting to decimals, we have .7 P ("P" is just my own shorthand here for "Passenger") and .3 NP ("NP" = "Not Passenger").

Step 4: *Repeat Step 3 for as many branches as you are given.*

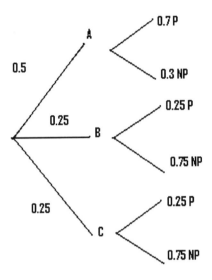

Step 5: *Multiply the probabilities of the first branch that produces the desired result together.* In our case, we want to know about the production of passenger planes, so we choose the first branch that leads to P.

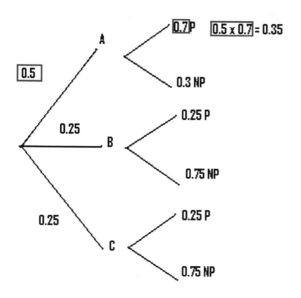

Step 6: *Multiply the remaining branches that produce the desired result.* In our example there are two more branches that lead to P.

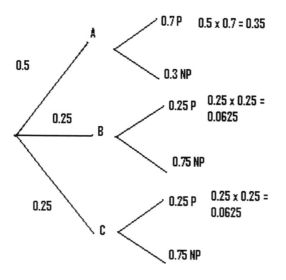

Step 7: *Add up all of the probabilities you calculated in Steps 5 and 6:* .35 + .625 + .625 = **.475**

HOW TO FIGURE OUT THE PROBABILITY OF PICKING FROM A DECK OF CARDS

Questions about how to figure out the odds of getting certain cards drawn from a deck are ubiquitous in basic statistics courses. A common question is the probability of choosing one card, and getting a certain number card (i.e. a 7) or one from a certain suit (i.e. a club).

Step 1: *figure out the total number of cards you might pull.* Write down all the possible cards and mark the ones that you would pull out. In our example we've been asked the probability of a club *or* a 7 so we're going to mark all the clubs and all the sevens:

hearts: 2, 3, 4, 5, 6, **7**, 8, 9, 10, J, Q, K, A
clubs: **2, 3, 4, 5, 6, 7, 8, 9, 10, J, Q, K, A**
spades: 2, 3, 4, 5, 6, **7**, 8, 9, 10, J, Q, K, A
diamonds: 2, 3, 4, 5, 6, **7**, 8, 9, 10, J, Q, K, A

This totals 16 cards.*

Step 2: *Count the total number of cards in the deck(s).* We have one deck, so the total is **52**.

Step 3: *Express the result as a fraction.* Divide **Step 3** by **Step 4:**

$$16 / 52$$

* Be careful not to double count the 7 that is also a club. It would be easy to make the mistake of saying that there are four 7s in the deck, and 13 clubs, therefore the total is 17. But that's wrong because one of the 7s was already counted!

HOW TO FIGURE OUT MUTUALLY EXCLUSIVE EVENTS

Often in elementary statistics, you'll be asked to figure out if events are mutually exclusive.

Sample Problem: If P(A) = .20, P(B) =.35, and (A U B) = .51, then are A and B mutually exclusive?

You would read that as: "If the Probability of A is 20%, the probability of B is 35%, and the probability of A given that B is true is 51%, then are A and B mutually exclusive?"

Step 1: *Add up the separate events (A and B)*:

$$.20 + .35 = .55$$

Step 2: *Compare your answer to the given "union" statement (A U B).* If they are the same, the events are mutually exclusive. If they are different, they are not mutually exclusive.

In our example, .55 does not equal .51, so the events are **not mutually exclusive**.

HOW TO TELL IF AN EVENT IS DEPENDENT OR INDEPENDENT

Being able to tell the difference between a dependent and independent event is vitally important in solving probability questions.

Step 1: *Is it possible for the events to occur in any order?* If no (the steps must be performed in a certain order), go to Step 3a. If yes (the Steps can be performed in any order), go to Step 2. If you are unsure, go to Step 2.

Some examples of events that can clearly be performed in any order are:

- Tossing a coin, then rolling a die
- Purchasing a car, then purchasing a coat
- Drawing cards from a deck

Some events that must be performed in a certain order are:

- Parking and getting a parking ticket (you can't get a parking ticket without parking)
- Surveying a group of people, and finding out how many women are against gun rights (because you are splitting the survey into subgroups, and you can't split a survey into subgroups without first performing the survey).

Step 2: *Ask yourself, does one event in any way affect the outcome (or the odds) of the other event?* If yes, go to Step 3a, if no, go to Step 3b.

Some examples of events that affect the odds or probability of the next event include:

- Choosing a card, not replacing it, then choosing another, because the odds of choosing the first card are 1/52, but if you do not replace it, you are changing the odds to 1/51 for the next draw.

- Choosing anything and not replacing it, then choosing another (i.e. choosing bingo balls, raffle tickets)

- Some examples of events that do not affect the odds or probability of the next event occurring are:

- Choosing a card and replacing it, then choosing another card (because the odds of choosing the first card are 1/52, and the odds of choosing the second card are 1/52)

- Choosing anything, as long as you put the items back

Step 3a: *You're done—the event is* **dependent.**

Step 3b: *You're done—the event is* **independent.**

DICE ROLLING

Dice rolling questions are common in probability and statistics. You might be asked the probability of rolling a five and a seven, or rolling a twelve. By far the easiest (visual) way to solve these types of problems is by writing out a **sample space**.

A sample space is the set of all possible probabilities. For example, in order to know what the odds are of rolling a 4 or a 7 from a set of two dice, we would first need to find out what all the possible combinations are. We could roll a double one [1][1], or a one and a two [1][2]. In fact, there are 36 possible combinations.

Step 1: Write out your sample space. For two dice, the **36** different possibilities are:

[1][1], [1][2], [1][3], [1][4], [1][5], [1][6], [2][1], [2][2], [2][3], [2][4], [2][5], [2][6], [3][1], [3][2], [3][3], [3][4], [3][5], [3][6], [4][1], [4][2], [4][3], [4][4], [4][5], [4][6], [5][1], [5][2], [5][3], [5][4], [5][5], [5][6], [6][1], [6][2], [6][3], [6][4], [6][5], [6][6]

Step 2: Look at your sample space and find how many add up to 4 or 7 (because we're looking for the probability of rolling one of those numbers).

[1][1], [1][2], **[1][3]**, [1][4], [1][5], **[1][6]**, [2][1], **[2][2]**, [2][3], [2][4],**[2][5]**, [2][6], **[3][1]**, [3][2], [3][3], **[3][4]**, [3][5], [3][6], [4][1], [4][2], **[4][3]**, [4][4], [4][5], [4][6], [5][1], **[5][2]**, [5][3], [5][4], [5][5], [5][6], **[6][1]**, [6][2], [6][3], [6][4], [6][5], [6][6].

There are **9** possible combinations.

Step 3: Take the answer from Step 2, and divide it by the size of your total sample space from Step l:

9 /36 = .25 You're done!

HOW TO CALCULATE BAYES THEOREM

$$P(A|B) = \frac{P(B|A)P(A)}{P(B)}$$

The formal definition of Bayes' Rule.

Bayes Theorem is a way to find conditional probability. In other words, it's a way to calculate the probability of an event happening if another event has already happened. For example, you might be interested in finding out a person's probability of catching norovirus if they live in a nursing home. "A" in the formula might mean the event "Person lives in a nursing home" and "B" might mean "Person has norovirus." Bayes theorem is pretty easy to calculate. The hard part is extracting the information from a word problem.

How to Calculate Bayes Theorem: Steps.

Sample Problem: In a particular pain clinic, 10% of patients are prescribed narcotic pain killers. Overall, five percent of the clinic's patients are addicted to narcotics (including pain killers and illegal substances). Among the addicts in the clinic, 80% have been prescribed narcotics. *If a patient is prescribed pain pills, what is the probability that they will become an addict?*

Step 1: Figure out what your event "A" is from the question. That information is in the italicized part of this particular question. The event that happens first (A) is being prescribed pain pills. That's given as 10%.

Step 2: Figure out what your event "B" is from the question. That information is also in the italicized part of this particular question.

Event B is being an addict. That's given as 5%.

Step 3: Figure out what the probability of event B (Step 2) given event A (Step 1). In other words, find what (B|A) is. In this question, we want to know "What is the probability of an addict being prescribed pain pills?" That is also given in the question as 80%.

Step 4: Insert your answers from Steps 1, 2 and 3 into the formula and solve.

$P(A|B) = P(B|A) * P(A) / P(B) = (0.8 * 0.1)/0.5 = 0.16$

If the patient is prescribed pain pills, their chance of becoming an addict is 0.16 (16%).

Binomial Probability Distributions

HOW TO DETERMINE IF AN EXPERIMENT IS BINOMIAL

Determining if a question concerns a binomial experiment involves asking four questions about the problem.

Sample Problem: which of the following are binomial experiments?

- Telephone surveying a group of 200 people to ask if they voted for George Bush.
- Counting the average number of dogs seen at a veterinarian's office daily.
- You are at a fair, playing "pop the balloon" with 6 darts. There are 20 balloons, and inside each balloon there's a ticket that says "win" or "lose."

Step 1: *Are there a fixed number of trials?*

Question 1: yes, there are 200 (maybe binomial)

Question 2: no (not binomial)

Question 3: yes, there are 6 (maybe binomial)

Step 2: *Are there only 2 possible outcomes?*

Question 1: yes, they did or didn't vote for Mr. Bush

Question 3: yes, you either win or lose

Step 3: *Are the outcomes independent of each other?* In other words, does the outcome of one trial (or one toss, or one question) affect another trial?

Question 1: yes, one person voting for Mr. Bush doesn't affect the next person's response.

Question 3: I would say that the outcomes are probably independent, although you could argue that for each balloon popped, it increases or decreases your chance of an individual win since there are fewer unpopped balloons. Which brings us to...

Step 4: *Does the probability of success remain the same for each trial?*

Question 1: yes, each person has the same chance of voting for Mr. Bush as the last.

Question 3: yes, no matter how many balloons you pop, each has a 50% chance of being a winner.

HOW TO FIND THE MEAN: PROBABILITY DISTRIBUTION OR BINOMIAL DISTRIBUTION

Sample Problem: A grocery store has determined that in crates of tomatoes, 95% carry no rotten tomatoes, 2% carry one rotten tomato, 2% carry two rotten tomatoes, and 1% carry three rotten tomatoes. Find the mean for the rotten tomatoes.

Step 1: *Convert all the percentages to decimal probabilities.*

95% = **.95**, 2% = **.02**, 1% = **.01**

Step 2: *Construct a probability distribution table.* (If you don't know how to do this, see the article on *constructing a probability distribution*).

Rotten Tomatoes X	0	1	2	3
Probability P(X)	.95	.02	.02	.01

Step 3: *Multiply the values in each column. In other words, multiply each value of X by each probability P(X).*
Referring to our probability distribution table:

$0 \times .95 = 0$
$1 \times .02 = .02$
$2 \times .02 = .04$
$3 \times .01 = .03$

Step 4: *Add the results from Step 3 together.*

$0 + .02 + .04 + .03 = $ **.09**

HOW TO FIND AN EXPECTED VALUE FOR A SINGLE ITEM

Sample Problem: You buy one $10 raffle ticket for a new car valued at $15,000. Two thousand tickets are sold. What is the expected value of the raffle?

Step 1: *Construct a probability chart.* Put Gain(X) and Probability P(X) heading the rows and Win/Lose heading the columns.

	Win	Lose
Gain X		
Probability P(X)		

Step 2: *Figure out how much you could gain and lose.* In our example, if we won, we'd be up $15,000 less the $10 cost of the raffle ticket. If you lose, you'd be down $10. Fill in the data.

	Win	Lose
Gain X	$14,990	-$10
Probability P(X)		

Step 3: *In the bottom row, put your odds of winning or losing.* 2,000 tickets were sold, so you have a 1 in 2,000 chance of winning, and you also have a 1,999 in 2,000 chance of losing.

	Win	Lose
Gain X	$14,990	-$10
Probability P(X)	1/2,000	1,999/2,000

Step 4: *Multiply the gains (X) in the top won by the Probabilities (P) in the bottom row.*

$14,990 × 1/2,000 = **$7.495**
-$10 × 1,999/2,000 = **-$9.995**
Step 5: *Add the numbers from Step 4 together to get the total expected value.*

$7.495 + -$9.995 = **-$2.50**

In other words, when entering this raffle we can expect to lose $2.50, so it's not a wise idea to enter.

HOW TO FIND AN EXPECTED VALUE FOR MULTIPLE ITEMS

Sample Problem: You buy one $10 raffle ticket, for the chance to win a car valued at $15,000, a CD player valued at $110, or a luggage set valued at $210. If 200 tickets are sold, what is the expected value of the raffle?

Step 1: *Construct a probability chart as in the "How to Find an Expected Value for a Single Item," except add columns for each possible prize:*

	Win	Win	Win	Lose
	Car	CD	Luggage	Cost
Gain X	$14,990	$100	$200	-$10
Probability P(X)	1/200	1/200	1/200	197/200

Step 2: *Multiply the gains (X) in the top won by the Probabilities (P) in the bottom row for each column:*

$14,990 × 1/200 = **$74.95**

$100 × 1/200 = **$.50**

$200 × 1/200 = **$1.00**

-$10 × 197/200 = **-$9.85**

Step 3: *Add the numbers from Step 2 together to get the total expected value.* $74.95 + $.50 + $1.00 + -$9.85 = **$66.60**

In other words, when entering this raffle, we can expect to gain $66.60, so it makes sense to enter.

Note on the formula: The actual formula for expected gain in probability and statistics is

$$E(X) = \sum (X \times P(X))$$

What this says in English is "The expected value is the sum of all the gains multiplied by their individual probabilities."

EXPECTED VALUE DISCRETE RANDOM VARIABLE: HOW TO FIND IT

Expected Value Discrete Random Variable: What is it?

You can think of an expected value as a mean, or average, for a probability distribution. A discrete random variable is a random variable that can only take on a certain number of values. For example, if you were rolling a die, it can only have the set of numbers {1,2,3,4,5,6}. The expected value formula for a discrete random variable is:

$$E(X) = \sum_{x \in \Omega} xm(x)$$

Basically, all the formula is telling you to do is find the mean by adding the probabilities. The mean and the expected value are so closely related they are basically the same thing. You'll need to do this slightly differently depending on if you have a set of values, a set of probabilities, or a formula.

Expected Value Discrete Random Variable (given a list).

Sample problem #1: The weights (X) of patients at a clinic (in pounds), are: 108, 110, 123, 134, 135, 145, 167, 187, 199. Assume one of the patients is chosen at random. What is the expected value?

Step 1: Find the mean. The mean is:

108 + 110 + 123 + 134 + 135 + 145 + 167 + 187 + 199 = 145.333.

That's it!

Expected Value Discrete Random Variable (given "X").

Sample problem #2. You toss a fair coin three times. X is the number of heads which appear. What is the expected value?
Step 1: Figure out the possible values for X. For a three coin toss, you could get anywhere from 0 to 3 heads. So your values for X are 0,1,2 and 3.

Step 2: Figure out your probability of getting each value of X. You may need to use a sample space. The probabilities are: 1/8 for 0 heads, 3/8 for 1 head, 3/8 for two heads, and 1/8 for 3 heads.

Step 3: Multiply your X values in Step 1 by the probabilities from step 2.
$E(X) = 0(1/8) + 1(3/8) + 2(3/8) + 3(1/8) = 3/2$.

The expected value is 3/2.

Expected Value Discrete Random Variable (given a formula, f(x)).

Sample problem #3. You toss a coin until a tail comes up. The probability distribution function is $f(x) = 1/2^x$. What is the expected value?
Step 1: Insert your "x" values into the first few values for the formula, one by one. For this particular formula, you'll get:
$1/2^0 + 1/2^1 + 1/2^2 + 1/2^3 + 1/2^4 + 1/2^5$

Step 2: Add up the values from **Step 1**:
$= 1 + 1/2 + 1/4 + 1/8 + 1/16 + 1/32 = 1.96875$.

Note: What you are looking for here is a number that the series converges on (i.e. a set number that the values are heading towards). In this case, **the values are headed towards 2, so that is**

your expected value.

Tip: You can only use the expected value discrete random variable formula if your function converges absolutely. In other words, the function must stop at a particular value. If it doesn't converge, then there is no expected value.

STANDARD DEVIATION: BINOMIAL DISTRIBUTION

Most textbooks in elementary statistics will ask you to find the standard deviation in the binomial probability distribution chapter. It's time consuming because the variance and standard deviation formulas require a lot of arithmetic, but it's not difficult at all!

Sample Problem: A sausage making machine occasionally produces misshapen sausages. The odds of the machine producing 0, 1, 2, 3, 4, or 5 sausages are .09, .07, .1, .04, .12 and .02 respectively. Find the variance and standard deviation.

Step 1: *Find the mean.*

Step 2: *Square the mean from Step 1.* For example, if your mean is .9, $.9^2$ = **.81**. Set this number aside.

Step 3: *Make a probability distribution chart.*

Number of misshapen sausages X	0	1	2	3	4	5
Probability P(X)	.09	.07	.1	.04	.12	.02

Step 4: *Square the top number, X, in each column and multiply it by the bottom number in the column, P(X). For example $0^2 \times .09$ from the first column, $1^2 \times .07$ from the second column. Repeat for all columns.*

Step 5: *Add all of the numbers in Step 4 together:* **0+.07+.4+.36+1.92+0.5=3.25**

Step 6: *Subtract the number you found in Step 2 from the number you found in Step 5.*

Step 5 – Step 2 = 3.25 - .81 = 2.44

Step 7: *Take the square root of the number you found in Step 6.*

√*Step 6 = 1.562*
That's it!

HOW TO WORK A BINOMIAL DISTRIBUTION FORMULA

The binomial formula can calculate the probability of success for binomial distributions. Often you'll be told to "plug in" the numbers to the formula and calculate. If you have a Ti-83 or Ti-89, the calculator can do much of the work for you. If not, here's how to break down the problem into simple steps so you get the answer right every time. Note that ! is a factorial (see the section on factorials if you don't know what that is).

Sample Problem: 80% of people who purchase pet insurance are women. If 9 pet insurance owners are randomly selected, find the probability that exactly 6 are women.

$$P(X) = \frac{n!}{(n-X)!\,X!} * p^x * q^{n-x}$$

Step 1: *Identify 'n' and 'X' from the problem.* Using our sample problem, n (the number of randomly selected items) is **9**, and X (the number you are asked to find the probability for) is **6**.

Step 2: *Work the first part of the formula:*

n! / (n - X)! X!

= 9! / ((9 - 6)! × 6!)

= **84**

Step 3: *Find p and q.* P is the probability of success and q is the probability of failure. We are given p: 80%, or **.8**. So the probability of failure is:

1 − .8 = **.2**

Or 20%.

Step 4: *Work the second part of the formula:*

p^X

$= .8^6$

$= .262144$

Step 5: *Work the third part of the formula:*

q^{n-X}

$= 2^{9-6}$

$= .2^3$

$= .008$

Step 7: *Multiply your answer from Steps 3, 5, and 6 together:*

84 × .262144 × .008 = **.176**

In other words, there is a 17.6% chance that exactly 6 of the respondents will be female.

HOW TO USE THE CONTINUITY CORRECTION FACTOR

A continuity correction factor is used when you use a continuous function to approximate a discrete one; when you use a normal distribution table to approximate a binomial, you're going to have to use a continuity correction factor. It's as simple as adding or subtracting .5: use the following table to decide whether to add or subtract.

Given	In English, the probability that...	Correction
P (X = n)	X is exactly n	P (n − .5 < X < n + .5)
P (X > n)	X is greater than n	P(X > n + .5)
P (X ≥ n)	X is greater than or equal to n	P(X > n − .5)
P (X ≤ n)	X is less than or equal to n	P(X < n + .5)
P (X < n)	X is less than n	P(X < n − .5)

Step 1: *Identify the probability from your function.* Let's say you are given:

P (X ≥ 351)

Step 2: *Replace the number with n:*

P (X ≥ 351)

P (X ≥ n)

Step 3: *Find the answer from Step 2 in the above table:* Given P(X ≥ n), use **P (X ≥ n − .5)**

Step 4: *Put your number back into the right-hand equation:* P (X > n − .5)

P (X ≥ 351− .5)

Step 5: *Add or Subtract according to the equation:*

P (X ≥ 351-.5)

P (X ≥ 350.5)

HOW TO USE THE NORMAL APPROXIMATION TO SOLVE A BINOMIAL PROBLEM

When (n × p) and (n × q) are greater than 5, you can use the normal approximation to solve a binomial distribution problem. This article shows you how to solve those types of problem using the continuity correction factor.

Sample Problem: Sixty two percent of 12th graders attend school in a particular urban school district. If a sample of 500 12th grade children is selected from the district, find the probability that at least 290 are actually enrolled in that school.

Step 1: *Determine p, q, and n:*

p is defined in the question as 62%, or **.62**
q is 1 – p: 1 -.62 = **.38**
n is defined in the question as **500**

Step 2: *Determine if you can use the normal distribution:*

n × p= 500 × .62 = **310**

n × q= 500 × .38 = **190**

These are both larger than 5, so we can use the normal approximation.

Step 3: *Find the mean, μ by multiplying n and p:*

n × p= **310**

Step 4: *Multiply the mean from Step 3 by q:*

310 × .38 = **117.8**

Step 5: *Take the square root of Step 4 to get the standard deviation, σ:*

$\sqrt{117.8} = \mathbf{10.85}$

Step 6: *Write the problem using correct notation*:

$P (X \geq 290)$

Step 7: *Rewrite the problem using the* continuity correction factor. To learn how, see the section above, titled "How to Use the Continuity Correction Factor":

$P (X \geq 290)$

$P (X \geq 290 - .5) = \mathbf{P (X \geq 289.5)}$

Step 8: *Draw a diagram with the mean in the center and your corrected n value from Step 7.* Shade the area that corresponds to the probability you are looking for. We're looking for X > 289.5:

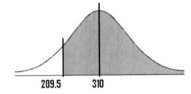

Step 9: *Find the z-value.* You can find this by subtracting the mean, μ, from the n value you found in Step 7, then dividing by the standard deviation, σ:

$(289.5 - 310) / 10.85 = \mathbf{-1.89}$

Step 10: *Look up the z-value in the z-table (see Resources section at the back of this book):*

The area for -1.819 is **.4706**

Step 11: *Add .5 to your answer in Step 10 to find the total area pictured:*

.4706 + .5 = **.9706**
That's it! The probability is .9706, or **97.06%**.

Normal Distributions

THE Z SCORE FORMULA: ONE SAMPLE

The basic z score formula for finding a z-score for a sample is:

$z = x - \mu / \sigma$

For example, let's say you have a test score of 190. The test has a mean (μ) of 150 and a standard deviation (σ) of 25. Assuming a normal distribution, your z score would be:

$z = x - \mu / \sigma$

$= 190 - 150 / 25 = 1.6$.

The z score tells you how many standard deviations from the mean your score is. In this example, your score is 1.6 standard deviations *above* the mean (because the score is positive).

$$z_i = \frac{x_i - \bar{\bar{x}}}{s}$$

Alternate form of the z score.

You may also see the z score formula shown above. This is exactly the same formula as $z = \bar{x} - \mu / \sigma$, except that $\bar{x}$ (the sample mean) is used instead of μ (the population mean) and s (the sample standard deviation) is used instead of σ (the population standard deviation). The steps for solving it are exactly the same.

Z Score Formula: Standard Error of the Mean

When you have multiple samples and want to describe the standard deviation of those sample means (the standard error), you would use this z score formula:

$z = \bar{x} - \mu / (\sigma / \sqrt{n})$

This z score will tell you **how many standard errors there are between the sample mean ($\bar{x}$) and the population mean (μ).**

Sample Problem: In general, the mean height of women (μ) is 65" with a standard deviation (σ) of 3.5". What is the probability of finding a random sample of 50 women (n) with a mean height ($\bar{x}$) of 70", assuming the heights are normally distributed?

$z = \bar{x} - \mu / (\sigma / \sqrt{n})$
$= 70 - 65 / (3.5/\sqrt{50}) = 5 / 0.495 = 10.1$

That's 10.1 standard deviations away from the mean.

We know that 99% of values fall within 3 standard deviations from the mean in a normal probability distribution. Therefore, there's less than 1% probability that any sample of women will have a mean height of 70".

HOW TO FIND THE AREA UNDER A NORMAL DISTRIBUTION CURVE BETWEEN 0 AND A GIVEN Z VALUE

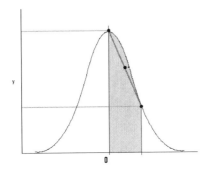

Sample Problem: Find the area between 0 and .46.
Step 1: *Look in the z-table for the given z-score by finding the intersection.* See the Tables section in the back of the book for a link to an online z-table. In our example, we look up .46*.

0.4 is in the left hand column and .06 in the top row. The intersection is **.1772**, which is the answer.

z	0.00	0.01	0.02	0.03	0.04	0.05	0.06
0.0	0.0000	0.0040	0.0080	0.0120	0.0160	0.0199	0.0239
0.1	0.0398	0.0438	0.0478	0.0517	0.0557	0.0596	0.0636
0.2	0.0793	0.0832	0.0871	0.0910	0.0948	0.0987	0.1026
0.3	0.1179	0.1217	0.1255	0.1293	0.1331	0.1368	0.1406
0.4	0.1554	0.1591	0.1628	0.1664	0.1700	0.1736	0.1772
0.5	0.1915	0.1950	0.1985	0.2019	0.2054	0.2088	0.2123

* **Note**: Because the graphs are symmetrical, you can ignore the negative z-scores and just look up their positive counterparts. For example, if you are asked for the area of 0 to -.46, just look up .46.

HOW TO FIND THE AREA UNDER THE NORMAL DISTRIBUTION CURVE IN ANY TAIL

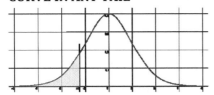

Sample Problem: Find the area in the tail to the left of z = -.46.

Step 1: *Look in the z-table (see back of book for link) for the given z- value by finding the intersection.* In our example, we look up .46*: .4 in the left hand column and .06 in the top row. The intersection is
.1772.

z	0.00	0.01	0.02	0.03	0.04	0.05	0.06
0.0	0.0000	0.0040	0.0080	0.0120	0.0160	0.0199	0.0239
0.1	0.0398	0.0438	0.0478	0.0517	0.0557	0.0596	0.0636
0.2	0.0793	0.0832	0.0871	0.0910	0.0948	0.0987	0.1026
0.3	0.1179	0.1217	0.1255	0.1293	0.1331	0.1368	0.1406
0.4	0.1554	0.1591	0.1628	0.1664	0.1700	0.1736	0.1772
0.5	0.1915	0.1950	0.1985	0.2019	0.2054	0.2088	0.2123

Step 2: *Subtract the z-value you just found in Step 1 from .5:*

.5 - .1772 = **.3228**

* **Note**: Because the graphs are symmetrical, you can ignore the negative z-values and just look up their positive counterparts. For example, if you are asked for the area of a tail on the left to -.46, just look up .46.

HOW TO FIND THE AREA UNDER THE NORMAL DISTRIBUTION CURVE BETWEEN TWO Z-SCORES ON ONE SIDE OF THE MEAN

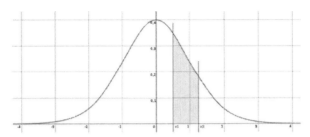

Sample Problem: Find the area from z = -.46 to z = -.40.

Step 1: *Look in the z-table (see back of book for resources) for the given z-scores (you should have two) by finding the intersections.* In our example, we look up both .40* and .46*: .4 in the left hand column and .06 in the top row; the intersection is **.1772**:.4 in the left hand column and .00 in the top row that intersection is **.1554**.

z	0.00	0.01	0.02	0.03	0.04	0.05	0.06
0.0	0.0000	0.0040	0.0080	0.0120	0.0160	0.0199	0.0239
0.1	0.0398	0.0438	0.0478	0.0517	0.0557	0.0596	0.0636
0.2	0.0793	0.0832	0.0871	0.0910	0.0948	0.0987	0.1026
0.3	0.1179	0.1217	0.1255	0.1293	0.1331	0.1368	0.1406
0.4	0.1554	0.1591	0.1628	0.1664	0.1700	0.1736	0.1772
0.5	0.1915	0.1950	0.1985	0.2019	0.2054	0.2088	0.2123

Step 2: *Subtract the smaller z-value you just found in Step 1 from the larger value.*

.1772 - .1554 = **.0218**

* **Note:** Because the graphs are symmetrical, ignore the negative z-values and look up their positive counterparts. For example, if you are asked for the area of a tail on the left

to -.46, just look up .46.

HOW TO FIND THE AREA UNDER THE NORMAL DISTRIBUTION CURVE BETWEEN TWO Z SCORES ON OPPOSITE SIDES OF THE MEAN

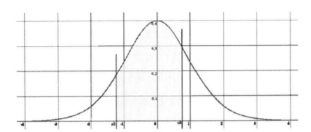

Sample Problem: Find the area from z = -.46 to z = .16.

Step 1: *Look in the z-table* (see back of book for resources) *for the given z-scores (you should have two) by finding the intersections.* In our example, we, look up both .46* and .06.

.4 in the left hand column and .06 in the top row; the intersection is **.1772**:.0 in the left hand column, and .06 in the top row; the intersection is **.0239**.

z	0.00	0.01	0.02	0.03	0.04	0.05	0.06
0.0	0.0000	0.0040	0.0080	0.0120	0.0160	0.0199	0.0239
0.1	0.0398	0.0438	0.0478	0.0517	0.0557	0.0596	0.0636
0.2	0.0793	0.0832	0.0871	0.0910	0.0948	0.0987	0.1026
0.3	0.1179	0.1217	0.1255	0.1293	0.1331	0.1368	0.1406
0.4	0.1554	0.1591	0.1628	0.1664	0.1700	0.1736	0.1772
0.5	0.1915	0.1950	0.1985	0.2019	0.2054	0.2088	0.2123

Step 2: *Add both of the values you found in Step 1 together.*

.1772 + .0239 = **.2011**

***Note:** Because the graphs are symmetrical, ignore the negative z-values and look up their positive counterparts.

HOW TO FIND THE AREA UNDER THE NORMAL DISTRIBUTION CURVE LEFT OF A Z SCORE

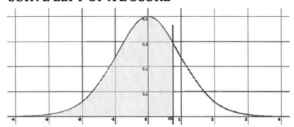

Sample Problem: Find the area to the left of the z value .46.
Step 1: *Look in the z-table (see back of book for resources) for the given z-value by finding the intersection.* In our example, we look up.46*: .4 in the left hand column and .06 in the top row. The intersection is **.1772.**

z	0.00	0.01	0.02	0.03	0.04	0.05	0.06
0.0	0.0000	0.0040	0.0080	0.0120	0.0160	0.0199	0.0239
0.1	0.0398	0.0438	0.0478	0.0517	0.0557	0.0596	0.0636
0.2	0.0793	0.0832	0.0871	0.0910	0.0948	0.0987	0.1026
0.3	0.1179	0.1217	0.1255	0.1293	0.1331	0.1368	0.1406
0.4	0.1554	0.1591	0.1628	0.1664	0.1700	0.1736	0.1772
0.5	0.1915	0.1950	0.1985	0.2019	0.2054	0.2088	0.2123

Step 2: *Add .5 to the z-value you just found in* **Step 1:**

*.5 + .1772 = **.6772***

* **Note:** Because the graphs are symmetrical, you can ignore the negative z-values and just look up their positive counterparts. For example, if you are asked for the area of a tail on the left to -.46, just look up .46.

HOW TO FIND THE AREA UNDER THE NORMAL DISTRIBUTION CURVE TO THE RIGHT OF A Z SCORE

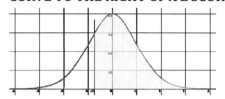

Sample Problem: Find the area to the right of the z value -.46.
Step 1: *Look in the z-table for the given z-value by finding the intersection.* In our example, we look up .46*.

.4 in the left hand column and .06 in the top row. The intersection is **.1772**.

z	0.00	0.01	0.02	0.03	0.04	0.05	0.06
0.0	0.0000	0.0040	0.0080	0.0120	0.0160	0.0199	0.0239
0.1	0.0398	0.0438	0.0478	0.0517	0.0557	0.0596	0.0636
0.2	0.0793	0.0832	0.0871	0.0910	0.0948	0.0987	0.1026
0.3	0.1179	0.1217	0.1255	0.1293	0.1331	0.1368	0.1406
0.4	0.1554	0.1591	0.1628	0.1664	0.1700	0.1736	0.1772
0.5	0.1915	0.1950	0.1985	0.2019	0.2054	0.2088	0.2123

Step 2: Add .5 to the *z-value you just found in* **Step 1:**

.5 + .1772 = **.6772**

* **Note**: Because the graphs are symmetrical, you can ignore the negative z-values and just look up their positive counterparts. For example, if you are asked for the area of a tail on the left to -.46, just look up .46.

HOW TO FIND THE AREA UNDER THE NORMAL DISTRIBUTION CURVE OUTSIDE OF A RANGE (TWO TAILS)

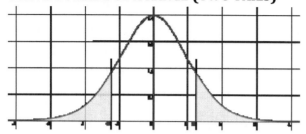

Sample Problem: Find the area to the left of the z value -.46 and to the right of z value .46.

Step 1: *Look in the z-table for one of the given z-values by finding the intersection.* In our example, look up .46*.

.4 in the left hand column and .06 in the top row. The intersection is **.1772**.

Z	0.00	0.01	0.02	0.03	0.04	0.05	0.06
0.0	0.0000	0.0040	0.0080	0.0120	0.0160	0.0199	0.0239
0.1	0.0398	0.0438	0.0478	0.0517	0.0557	0.0596	0.0636
0.2	0.0793	0.0832	0.0871	0.0910	0.0948	0.0987	0.1026
0.3	0.1179	0.1217	0.1255	0.1293	0.1331	0.1368	0.1406
0.4	0.1554	0.1591	0.1628	0.1664	0.1700	0.1736	0.1772
0.5	0.1915	0.1950	0.1985	0.2019	0.2054	0.2088	0.2123

Step 2: *Subtract the z-value you just found in Step 1 from .5:*

.5 - .1772 = **.3228**

Step 3: *Repeat Steps 1 and 2 for the other tail.* In our example, *the* *other tail also uses a z value of .46:*

.5 - .1772 = **.3228**

Step 4: *Add both z-values together.*

*.3228 + .3228 = **.6456***

* **Note**: Because the graphs are symmetrical, you can ignore the negative z-values and just look up their positive counterparts. For example, if you are asked for the area of a tail on the left to -.46, just look up .4

HOW TO CALCULATE CHEBYSHEV'S THEOREM

The formula is: $1-(1/k^2)$, where k is the number of standard deviations.

Sample Problem: A left-skewed distribution has a mean of 4.99 and a standard deviation of 3.13. Use Chebyshev's Theorem to calculate the proportion of observation you would expect to find within two standard deviations (in other words, between -2 and +2 standard deviations) from the mean:

Step 1: *Square the number of standard deviations*:
$2^2 = $ **4**

Step 2: *Divide 1 by your answer to **Step 1**:*
1 / 4 = **0.25**

Step 3: *Subtract Step 2 from 1*:
1 − 0.25 = **0.75**

At least **75%** of the observations fall between -2 and +2 standard deviations from the mean.

Warning: As you may be able to tell, the mean of your distribution has no effect of Chebyshev's theorem! That fact can cause some wide variations in data, and some inaccurate results.

Central Limit Theorem

MEAN OF THE SAMPLING DISTRIBUTION OF THE MEAN

In a nutshell, the mean of the sampling distribution of the mean **is the same as the population mean**. For example, if your population mean (μ) is 99, then the mean of the sampling distribution of the mean, μ_m, is also 99 (as long as you have a sufficiently large sample size). Yes, it really is that simple! If you want to understand *why*, read on below.

The Central Limit Theorem.

Roughly stated, the central limit theorem tells us that if we have a large number of independent, identically distributed variables, the distribution will be approximately normal. It doesn't matter what the underlying distribution is.

Here's a simple example of the theory: when you roll a single die, your odds of getting any number (1,2,3,4,5, or 6) are the same (1/6). The mean for any roll is (1 + 2 + 3 + 4 + 5 + 6) / 6 = 3.5. The results from a one-die roll are shown in the following figure: it looks like a uniform distribution. However, as the sample size is increased (two dice, three dice...), the mean of the sampling distribution of the mean looks more and more like a normal distribution. That is what the central limit theorem predicts.

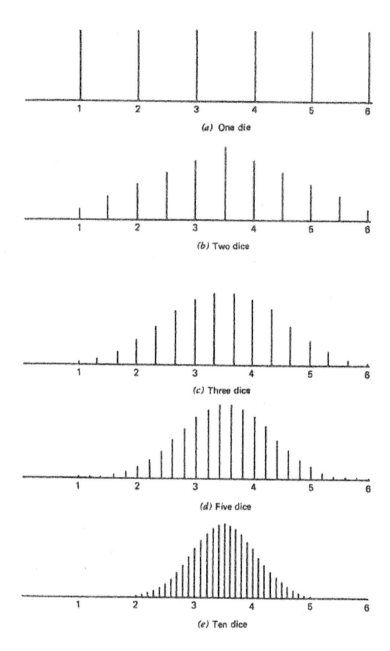

(a) One die

(b) Two dice

(c) Three dice

(d) Five dice

(e) Ten dice

Image: U of Michigan.

As the sample size increases, the mean of the sampling distribution of the mean will approach the population mean of μ, **and** the variance will approach σ^2/N, where N is the sample size.

You can think of a sampling distribution as a relative frequency distribution with a large number of samples.

The mean of the sampling distribution of the mean formula

The mean of the sampling distribution of the mean formula is $\mu_M = \mu$, where μ_M is the mean of the sampling distribution of the mean.

VARIANCE OF THE SAMPLING DISTRIBUTION OF THE SAMPLE MEAN

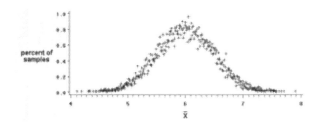

A sampling distribution where the mean = 6. Image: U of Oklahoma

The sampling distribution of the sample mean is a probability distribution of all the sample means. Let's say you had 1,000 people, and you sampled 5 people at a time and calculated their average height. If you kept on taking samples (i.e. you repeated the sampling a thousand times), eventually the mean of all of your sample means will:

1.   Equal the population mean, μ
2.   Look like a normal distribution curve.

The **variance** of this probability distribution gives you an idea of how spread out your data is around the mean. The larger the sample size, the more closely the sample mean will represent the population mean. In other words, as N grows larger, the variance becomes smaller. Ideally, when the sample mean matches the population mean, the variance will equal zero.

The formula to find the variance of the sampling distribution of the mean is:

$\sigma^2_M = \sigma^2 / N$,

where:

σ^2_M = variance of the sampling distribution of the sample mean.
σ^2 = population variance.
N = your sample size.

How to find the Variance of the sampling distribution of the sample mean

Sample question: If a random sample of size 19 is drawn from a population distribution with standard deviation σ=20 then what will be the variance of the sampling distribution of the sample mean?

Step 1: Figure out the population variance. Variance is the standard deviation squared, so:
$\sigma^2 = 20^2 = 400$.

Step 2: Divide the variance by the number of items in the sample. This sample has 19 items, so:
400 / 19 = 21.05.

That's it!

CENTRAL LIMIT THEOREM PROBLEM INDEX

A **Central Limit Theorem** word problem will most likely contain the phrase "assume the variable is normally distributed," or one like it. You will be given:

- A population (i.e. 29-year-old males, seniors between 72 and 76, all registered vehicles, all cat owners)
- An average (i.e. 125 pounds, 24 hours, 15 years, $196.42)
- A standard deviation (i.e. 14.4lbs, 3 hours, 120 months, $15.74)
- A sample size (i.e. 15 males, 10 seniors, 79 cars, 100 owners)

Choose one of the following questions and go to that section number:

Go to "Greater than" on p. 120: I want to find the probability that the mean is **greater** than a certain number.

Go to "less than" on p. 122: I want to find the probability that the mean is **less** than a certain number.

Go to "Between" on p. 124: I want to find the probability that the mean is **between** a certain set of numbers either side of the mean.

CENTRAL LIMIT THEOREM "GREATER THAN" PROBABILITY

Sample Problem: There are 250 dogs at a dog show who weigh an average of 12 pounds, with a standard deviation of 8 pounds. If 4 dogs are chosen at random, what is the probability they have an average weight of more than 25 pounds?

Step 1: *Identify the parts of the problem.* Your question should state:

- mean (average or μ) standard deviation (σ) population size
- sample size (n)
- number associated with "greater than" $(\overline{X})$

Step 2: *Draw a graph.* Label the center with the mean. Shade the area to the right of $(\overline{X})$. This step is optional, but it may help you see what you are looking for.

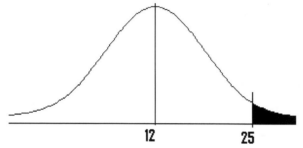

12   25

Step 3: *Use the following formula to find the z-value.* Plug in the numbers from Step 1.

$$z = \frac{\overline{X} - \mu}{\sigma/\sqrt{n}}$$

If formulas confuse you, all this formula is asking you to do is:

Subtract the mean (μ in Step 1) from the greater than value $(\overline{X}$ In Step 1). Set this number aside for a moment.

Divide the standard deviation (σ in Step 1) by the square root of your sample (n in Step 1).

Divide your result from Step 1 by your result from Step 2.

$$z = \frac{\bar{X} - \mu}{\sigma/\sqrt{n}}$$

$$z = \frac{25 - 12}{8/\sqrt{4}}$$

$$z = \frac{13}{8/\sqrt{4}}$$

$$z = \frac{13}{8/2}$$

$$z = \frac{13}{4}$$

$$z = 3.25$$

Step 4: *Look up the z-value you calculated in Step 3 in the* ***z-table*** *(see back of book for resources).*

Z value of 3.25 corresponds to ***.4494***

Step 5: *Subtract your z-score from .5:*

.5 - .4994 = ***.0006***

Step 6: *Convert the decimal in Step 5 to a percentage:*

.0006 = ***.06%***

CENTRAL LIMIT THEOREM "LESS THAN" PROBABILITY

Sample Problem: There are 250 dogs at a dog show who weigh an average of 12 pounds, with a standard deviation of 8 pounds. If 4 dogs are chosen at random, what is the probability they have an average weight of less than 25 pounds?

Step 1: *Identify the parts of the problem.* Your question should state:

- mean (average or μ)
- standard deviation (σ)
- population size
- sample size (n)
- number associated with "less than" $(\overline{X})$

Step 2: *Draw a graph.* Label the center with the mean. Shade the area to the left of "less than." This step is optional, but it may help you see what you are looking for.

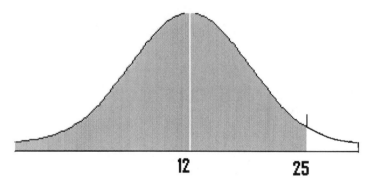

Step 3: *Use the following formula to find the z-value.* Plug in the numbers from Step 1.

$$z = (\overline{X} - \mu)/(\sigma/\sqrt{n})$$

If formulas confuse you, all this formula is asking you to do is:

Subtract the mean (μ in Step 1) from the less than value (in Step 1).

Set this number aside for a moment.

Divide the standard deviation (σ in Step 1) by the square root of your sample (n in Step 1).

Divide your result from Step 1 by your result from Step 2.

$$z = \frac{\bar{X} - \mu}{\sigma/\sqrt{n}}$$

$$z = \frac{25 - 12}{8/\sqrt{4}}$$

$$z = \frac{13}{8/\sqrt{4}}$$

$$z = \frac{13}{8/2}$$

$$z = \frac{13}{4}$$

$$z = 3.25$$

Step 4: *Look up the z-value you calculated in Step 3 in the z-table (see back of book for resources).*

*Z value of 3.25 corresponds to **.4494***

Step 5: *Add Step 4 to .5:*

*.5 + .4994 = **.9994***

Step 6: *Convert the decimal in Step 5 to a percentage:*

*.9994 = **99.94%***

CENTRAL LIMIT THEOREM "BETWEEN" PROBABILITY

Sample Problem: There are 250 dogs at a dog show who weigh an average of 12 pounds, with a standard deviation of 8 pounds. If 4 dogs are chosen at random, what is the probability they have an average weight of greater than 8 pounds and less than 25 pounds?

Step 1: *Identify the parts of the problem.* Your question should state:

- mean (average or μ) standard deviation (σ) population size
- sample size (n)
- number associated with "less than"
- number associated with "greater than"

Step 2: *Draw a graph.* Label the center with the mean. Shade the area between "less than" and "greater than". This step is optional, but it may help you see what you are looking for.

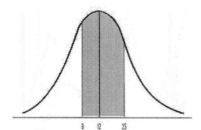

Step 3: *Use the following formula to find the z-values.* Plug the "less than" value in from **Step 1 as** $\bar{X}$:

$$z = \frac{\bar{X} - \mu}{\sigma/\sqrt{n}}$$

$$z = \frac{\bar{X} - \mu}{\sigma/\sqrt{n}}$$

$$z = \frac{25 - 12}{8/\sqrt{4}}$$

$$z = \frac{13}{8/\sqrt{4}}$$

$$z = \frac{13}{8/2}$$

$$z = \frac{13}{4}$$

$$z = 3.25$$

All this formula is asking you to do is:

a) Subtract the mean (μ in Step 1) from the greater than value in Step 1.
b) Divide the standard deviation (σ in Step 1) by the square root of your sample (n in Step 1).
c) Divide your result from *a* by your result from *b*.

Step 4: *Use the following formula to find the z-values.* Plug in the "less than" value from **Step 1**:

$$z = \frac{\bar{X} - \mu}{\sigma/\sqrt{n}}$$

$$z = \frac{8 - 12}{8/\sqrt{4}}$$

$$z = \frac{-4}{8/2}$$

$$z = \frac{-4}{4}$$

$$z = -1$$

Step 5: *Look up the z-value you calculated in Step 3 in the z-table (see back of book for resources).*

Z value of 3.25 corresponds to .4494

Step 6: *Look up the z-value you calculated in Step 4 in the z-table (see back of book for resources).*

Z value of 1 corresponds to .3413

Step 7: *Add Step 5 and 6 together:*

.4494 + .3413 = **.7907**

Step 8: *Convert the decimal in Step 7 to a percentage:*

.7907 = **79.07%**

Confidence Intervals and Sampling

HOW TO GET A STRATIFIED RANDOM SAMPLE

"Stratified" means "in layers," so in order to get a stratified random sample you first need to make the layers. What layers you have depends on characteristics of your population. For example, if you are surveying U.S. residents about their plans for retirement, you might want your layers to represent different age groups. The sample size for each strata (layer) is proportional to the size of the layer:

*Sample size of the strata = size of entire sample / population size * layer size.*

How to Get a Stratified Random Sample: Steps

Sample question: You work for a small company of 1,000 people and want to find out how they are saving for retirement. Use stratified random sampling to obtain your sample.

Step 1: Decide how you want to stratify (divide up) your population. For example, people in their twenties might have different saving strategies than people in their fifties.

Step 2: Make a table representing your strata. The following table shows age groups and how many people in the population are in that strata:

Age	Total Number of People in Strata
20-29	160
30-39	220
40-49	240
50-59	200
60+	180

Step 3: Decide on your sample size. If you don't know how to find a sample size, see: Sample size (how to find one). For this example, we'll assume your sample size is 50.

Step 4: Use the stratified sample formula (Sample size of the strata = size of entire sample / population size * layer size) to calculate the proportion of people from each group:

Age	Number of People in Strata	Number of People in Sample
20-29	160	50/1000 * 160 = 8
30-39	220	50/1000 * 220 = 11
40-49	240	50/1000 * 240 = 12
50-59	200	50/1000 * 200 = 10
60+	180	50/1000 * 180 = 9

Note that all of the individual results from the stratum add up to your sample size of 50: 8 + 11 + 12 + 10 + 9 = 50

Step 5: Perform random sampling (i.e. simple random sampling) in each stratum to select your survey participants.

That's how to get a stratified random sample!

Tip: Each element in your population should only fit into one stratum. In other words, one person cannot be in more than one group.

How to Perform Simple Random Sampling

Simple random sampling is the most basic sampling technique in statistics. A simple random sample is chosen in such a way that every set of individuals has an equal chance to be in the selected sample. It *sounds* easy, but SRS is often difficult to employ in surveys or experiments. In addition, it's very easy for bias to creep into samples obtained with simple random sampling. Sometimes it's impossible (either financially or time-wise) to get a realistic sampling frame (the population from which the sample is to be chosen). For example, if you wanted to study **all** the adults in the U.S. who had high cholesterol, the list would be practically impossible to get unless you surveyed every person in the country. Therefore other sampling methods would probably be better suited to that particular experiment.
The simplest example of SRS would be working with things like dice or cards — rolling the die or dealing cards from a deck can give you a simple random sample. But in real life you're usually dealing with people, not cards, and that can be a challenge.

How to Perform Simple Random Sampling: Example

Sample question: Outline the steps for obtaining a simple random sample for outcomes of strokes in U.S. trauma hospitals.

Step 1: Make a list of all the trauma hospitals in the U.S. (there are several hundred: the CDC keeps a list).

Step 2: Assign a sequential number to each trauma center (1,2,3...n). This is your sampling frame (the list from which you draw your simple random sample).

Step 3: Figure out what your sample size is going to be. See: (Sample size: how to find one).

Step 4: Use a random number generator to select the sample, using your sampling frame (population size) from Step 2 and your sample size from Step 3. For example, if your sample size is 50 and your population is 500, generate 50 random numbers between 1 and 500.

Warning: If you compromise (say, by not including ALL trauma centers in your sampling frame), it could open your results to bias.

HOW TO PERFORM SYSTEMATIC SAMPLING

When you're sampling from a population, you want to make sure you're getting a fair representation of that population. Otherwise, your statistics will be biased or skewed and perhaps meaningless. One way to get a fair and random sample is to assign a number to every population member choose the *nth* member from that population. For example, you could choose every 10th member, or every 100th member. This method of choosing the nth member is called **systematic sampling**.

Systematic sampling is quick and convenient when you have a complete list of the members of your population. However, if there's some kind of pattern to the original list, then bias may creep in to your statistics. For example, if a list of people is ordered as MFMFMFMF, then choosing every 10th number will give you a sample consisting entirely of females. How to perform systematic sampling without this type of sampling bias? You could randomly shuffle the list before choosing the nth item or you could use **repeated** systematic sampling. **Repeated systematic sampling** is a type of systematic sampling where you take several small samples from the same population. It's used if you aren't sure you have a completely random list and you want to avoid sample bias.

How to Perform Systematic Sampling: Steps

Step 1: Assign a number to every element in your population.

Step 2: Decide how large your sample size should be.
See: Sample size(how to find one).

Step 3: Divide the population by your sample size. For example, if your population is 100 and your sample size is 10, then:
100 / 10 = 10
This is your "nth" sampling digit (i.e. you'll choose every 10th item)
1 2 3 4 5 6 7 8 9 **10**
11 12 13 14 15 16 17 18 19 **20**
21 22 23 24 25 26 27 28 29 **30**
31 32 33 34 35 36 37 38 39 **40**
41 42 43 44 45 46 47 48 49 **50**
51 52 53 54 55 56 57 58 59 **60**
61 62 63 64 65 66 67 68 69 **70**
71 72 73 74 75 76 77 78 79 **80**
81 82 83 84 85 86 87 88 89 **90**
91 92 93 94 95 96 97 98 99 **100**

How to Perform Systematic Sampling: Repeated Systematic Sampling

Step 1: Assign a number to every element in your population.

Step 2: Decide how large your sample size should be. **See:** Sample size (How to find one).

Step 3: Divide the population by your sample size. For example, if your population is 100 and your sample size is 10, then:
100 / 10 = 10
This is your "nth" sampling digit (i.e. you'll choose every 10th item)

Step 4: Use the sampling digit from Step 3 up to a certain point. This is usually a judgment call. For this example, we'll sample up to 50.

1 2 3 4 5 6 7 8 9 **10**
11 12 13 14 15 16 17 18 19 **20**
21 22 23 24 25 26 27 28 29 **30**
31 32 33 34 35 36 37 38 39 **40**
41 42 43 44 45 46 47 48 49 **50**
51 52 53 54 55 56 57 58 59 60
61 62 63 64 65 66 67 68 69 70
71 72 73 74 75 76 77 78 79 80
81 82 83 84 85 86 87 88 89 90
91 92 93 94 95 96 97 98 99 100

Step 4: Switch to a different starting point and then continue sampling with the nth digit. Again, this is usually a judgment call. For this example, we'll switch from 50 to 51 as the new starting point (so 61 will be the first number).

1 2 3 4 5 6 7 8 9 **10**
11 12 13 14 15 16 17 18 19 **20**
21 22 23 24 25 26 27 28 29 **30**
31 32 33 34 35 36 37 38 39 **40**
41 42 43 44 45 46 47 48 49 **50**
51 52 53 54 55 56 57 58 59 60
61 62 63 64 65 66 67 68 69 70
71 72 73 74 75 76 77 78 79 80
81 82 83 84 85 86 87 88 89 90
91 92 93 94 95 96 97 98 99 100

Note that we only have 9 in our sample (we wanted 10), so **return to the beginning of the list** and continue:

1 2 3 4 5 6 7 8 9 10

How to Find a Sample Size in Statistics

Step 1: Conduct a census if you have a small population. A "small" population will depend on your budget and time constraints. For example, it may take a day to take a census of a student body at a small private university of 1,000 students but you may not have the time to survey 10,000 students at a large state university.

Step 2: Use a sample size from a similar study. Chances are, your type of study has already been undertaken by someone else. You'll need access to academic databases to search for a study (usually your school or college will have access). A pitfall: you'll be relying on someone else correctly calculating the sample size. Any errors they have made in their calculations will transfer over to your study.

Step 3: Use a table to find your sample size. If you have a fairly generic study, then there is probably a table for it. Ask your instructor, or perform a Google search for something like "95% confidence interval sample size table."

Step 4: Use a sample size calculator. You can find some online by searching for "sample size calculator".

Step 5: Use a formula. There are many different formulas you can use, depending on what you know (or don't know) about your population. If you know some parameters about your population (like a known standard deviation), you can use the techniques up next. If you don't know much about your population, use Slovin's formula.

HOW TO FIND A SAMPLE SIZE GIVEN A CONFIDENCE INTERVAL AND WIDTH (UNKNOWN STANDARD DEVIATION)

Sample Problem: 41% of Jacksonville residents said that they had been in a hurricane. How many adults should be surveyed to estimate the true proportion of adults who have been in a hurricane, with a 95% confidence interval 6% wide?

Step 1: *Using the data given in the question, figure out the following variables:*

- $z_{a/2}$: Divide the confidence interval by two, and look that area up in the z-table (see resources at back of book):
.95 / 2 = .475

The closest z-score for .475 is **1.96**

- E (margin of error): Divide the given width by 2.

6% / 2 = .06 / 2 = **.03**
- $\hat{p}$: use the given percentage.

41% = **.41**
- $\hat{q}$: subtract $\hat{p}$ from 1

1 - .41 = **.59**

Step 2: *Multiply $\hat{p}$ by $\hat{q}$. Set this number aside for a moment.*

.41 × .59 = **.2419**

Step 3: *Divide $z_{a/2}$ by E.*

1.96 / .03

= **65.3333333**

Step 4: *Square* ***Step 3:***

65.3333333 × 65.3333333

= **4,268.44444**

Step 5: *Multiply Step 2 by **Step 4:***

.2419 × 4268.44444

= 1,032.53671

= **1033** people to survey

HOW TO FIND A SAMPLE SIZE GIVEN A CONFIDENCE INTERVAL AND WIDTH (KNOWN STANDARD DEVIATION)

Sample Problem: Suppose we want to know the average age of the students on a particular college campus, plus or minus .5 years. We'd like to be 99% confident about our result. From a previous study, we know that the standard deviation for the population is
2.9 years.

Step 1: *Find z $_{a/2}$ by dividing the confidence interval by two, and looking that area up in the z-table (see resources in back of book):*

.99 / 2 = .495

The closest z-score for .495 is **2.58**

Step 2: *Multiply Step 1 by the standard deviation.*

2.58 × 2.9 = **7.482**

Step 3: *Divide Step 2 by the margin of error:*

*7.482 / .5 = **14.96***

Step 4: *Square **Step 3:***

$$14.96 \times 14.96 = \textbf{223.8016}$$

*So we need a sample of **224** people.*

SLOVIN'S FORMULA: WHAT IS IT AND WHEN DO I USE IT?

If you take a population sample, you usually use a formula to figure out what sample size you need to take. Sometimes you know something about a population, which can help you **determine a sample size**. For example, it's well known that IQ scores follow a normal distribution pattern. But what about if you know nothing about your population at all? That's when you can use Slovin's formula to figure out what sample size you need to take, which is written as n = N / (1 + Ne^2) where n = Number of samples, N = Total population and e = Error tolerance, or margin of error.

Sample question: Use Slovin's formula to find out what sample of a population of 1,000 people you need to take for a survey on their soda preferences.
Step 1: Figure out what you want your **confidence level** to be. For example, you might want a confidence level of 95 percent (which will give you a margin error of 0.05), or you might need better accuracy at the 98 percent confidence level (which produces a margin of error of 0.02). ME = 1 – Confidence interval.
Step 2. Plug your data into the formula. In this example, we'll use a 95 percent confidence level with a population size of 1,000.
n = N / (1 + N e^2) =
1,000 / (1 + 1000 * 0.05 ^{2}) = 285.714286
Step 3: Round your answer to a whole number (because you can't sample a fraction of a person or thing!)
285.714286 = 286

HOW TO FIND A CONFIDENCE INTERVAL FOR A POPULATION

Sample Problem: 510 people applied to the Bachelor's in Elementary Education program at a certain state college. Of those applicants, 57 were men. Find the 90% confidence interval of the true proportion of men who applied to the program.

Step 1: *Read the question carefully and figure out the following variables:*

- α: subtract the given confidence interval from 1.

1 - .9 = **.1**

- $z_{α/2}$: divide α by 2, then look up that area in the z-table.

.1 / 2 = .0500

The closest z-value to an area of .0500 is **0.13**

- $\hat{p}$: divide the proportion given (i.e. the smaller number) by the sample size.

57 / 510 = **.112**

- $\hat{q}$: Subtract $\hat{p}$ from 1.

1 - .112 = **.888**

Step 2: *Multiply $\hat{p}$ by $\hat{q}$:*

.112 × .888 = **.099456**

Step 3: *Divide Step 2 by the sample size:*

.099456 / 510 = **.000195011765**

Step 4: *Take the square root of Step 3:*

√.000195011765 = **.0139646613**

Step 5: *Multiply Step 4 by $z_{α/2}$:*

.0139646613 × 0.13 = **.0182**

Step 6: *For the lower percentage, subtract Step 5 from* $\hat{p}$:

.112 - .0182 = .0938 = **9.38%**

Step 7: *For the upper percentage, add Step 5 to* $\hat{p}$:

.112 + .0182 = .1302 = **13.02%**

The 90% confidence interval is between 9.38% and 13.02% of the population.

HOW TO FIND A CONFIDENCE INTERVAL FOR A SAMPLE

When you don't know anything about a population's behavior (i.e. you're just looking at data for a sample), you need to use the t-distribution table instead of the z-distribution table (see resources at the back of the book) to find the confidence interval.

Sample Problem: A group of 10 foot surgery patients had a mean weight of 240 pounds. The sample standard deviation was 25 pounds. Find the 95% confidence interval for the true mean weight of all foot surgery patients. Assume normal distribution.

Step 1: *Subtract 1 from your sample size.*

10 − 1 = **9**

This gives you degrees of freedom, which you'll need in Step 3.

Step 2: *Subtract the confidence level from 1, and then divide by two:*

(1 − .95) / 2 = **.025**

Step 3: *Look up your answers to Step 1 and 2 in the* t-distribution table. For 9 degrees of freedom (df) and α = .025, the result is **2.262**.

Df	α = 0.1	0.05	0.025	0.01	0.005	0.001	0.0005
∞	t_α=1.282	1.645	1.960	2.326	2.576	3.091	3.291
1	3.078	6.314	12.706	31.821	63.656	318.289	636.578
2	1.886	2.920	4.303	6.965	9.925	22.328	31.600
3	1.638	2.353	3.182	4.541	5.841	10.214	12.924
4	1.533	2.132	2.776	3.747	4.604	7.173	8.610
5	1.476	2.015	2.571	3.365	4.032	5.894	6.869
6	1.440	1.943	2.447	3.143	3.707	5.208	5.959
7	1.415	1.895	2.365	2.998	3.499	4.785	5.408
8	1.397	1.860	2.306	2.896	3.355	4.501	5.041
9	1.383	1.833	2.262				

Step 4: *Divide your sample standard deviation by the square root of your sample size:*

25 / √10 = **7.90569415**

Step 5: *Multiply Step 3 by* **Step 4:**

2.262 × 7.90569415 = **17.8826802**

Step 6: *For the lower end of the range, subtract Step 5 from the sample mean:*

240 – 17.8826802 = **222.117**

Step 7: *For the upper end of the range, add Step 5 to the sample mean:*

240 + 17.8826802 = **257.883**

HOW TO CONSTRUCT A CONFIDENCE INTERVAL FROM DATA USING THE T DISTRIBUTION

Sample Problem: Construct a 98% Confidence Interval for the mean based on the following data: 45, 55, 67, 45, 68, 79, 98, 87, 84, 82.

Step 1: *Find the mean and standard deviation for the data:*

Mean: **71**

Standard Deviation: **18.172**

For information on finding the mean, see, "How to Find the Mean, Mode, and Median." For information on finding the standard deviation, see, "How to Find the Sample Variance and Standard Deviation."

Step 2: *Subtract 1 from your sample size to find the degrees of freedom (df).* We have 10 numbers listed, so our sample size is 10:

10 – 1 = **9 degrees of freedom**

Step 3: *Subtract the confidence level from 1, and then divide by two:*

(1 - .98) / 2

= .02 / 2

= **.01**
This is your α level.

Step 4: *Look up df (Step 2) and α (Step 3) in the t-distribution table (see back of book for resources).* For df = 9 and α = .01, the table gives us **2.821**.

Step 4: *Divide your standard deviation (Step 1) by the square root of your sample size:*

18.172 / √10 = **5.75**

Step 5: *Multiply Step 3 by **Step 4:***

2.2821 × 5.75 = **13.122075**

Step 6: *For the lower end of the range, subtract Step 5 from the mean (Step 1):*

71 − 13.122075 = **57.877925**

Step 7: *For the upper end of the range, add Step 5 to the mean (Step 1):*

71 + 13.122075 = **84.122**

Our 98% confidence interval is **57.877925** to **84.122**.

FINDING CONFIDENCE INTERVALS FOR TWO POPULATIONS (PROPORTIONS)

Sample Problem: A study revealed that 65% of men surveyed supported the war in Afghanistan and 33% of women supported the war. If 100 men and 75 women were surveyed, find the 90% confidence interval for the data's true difference in proportions.

Step 1: *Find the following variables from the information given in the question:*

n_1 (population sample 1): **100**
$\hat{p}_1$ (population sample 1, positive response): 65% or **.65**
$\hat{q}_1$ (population sample 1, negative response): 35% or **.35**
n_2 (population sample 2): **75**
$\hat{p}_2$ (population sample 2, positive response): 33% or **.33**
$\hat{q}_2$ (population sample 2, negative response): 67% or **.67**

Step 2: *Find $z_{\alpha/2}$.* Subtract the confidence interval from 1, divide by two, and look that area up in the z-table (see resources at back of book):

$(1-.9) / 2 = .05$; $z_{\alpha/2} = $ **0.13**

Step 3: *Enter your data into the following formula and solve:*

$$(\hat{p}_1 - \hat{p}_2) - z_{\alpha/2}\sqrt{\frac{\hat{p}_1\hat{q}_1}{n_1} + \frac{\hat{p}_2\hat{q}_2}{n_2}} < p_1 - p_2 < (\hat{p}_1 - \hat{p}_2) + z_{\alpha/2}\sqrt{\frac{\hat{p}_1\hat{q}_1}{n_1} + \frac{\hat{p}_2\hat{q}_2}{n_2}}$$

If solving formulas scare you, here's how to solve that formula step-by-step:

1. Multiply $\hat{p}_1$ and $\hat{q}_1$ together.

.65 × .35 = **.2275**

2. Divide your answer to (2) by n_1. Set this number aside:
.2275 / 100 = **.00275**

3. Multiply $\hat{p}_2$ and $\hat{q}_2$ together:
.33 × .67 = **.2211**

4. Divide your answer to (3) by n_2:
.2211 / 75 = **.002948**

5. Add (3) and (4) together:
.00275 + .002948 = **.005698**

6. Take the square root of (5):
√.005698 = **.075485**

7. Multiply (6) by $z_{\alpha/2}$ found in **Step 2:**
.075485 × 0.13 = **.0098**

8. Subtract $\hat{p}_1$ from $\hat{p}_2$:
.65 - .33 = **.32**

9. Subtract (8) from (7) to get the left limit:
.32 - .0098 = **.319902**

10. Add (8) to (7) to get the right limit:
.32 + .0098 = **.320098**

Our 90% confidence interval is **.319902** to **.320098 (31.9902% to 32.0098%)**.
Easy!

Hypothesis Testing

HYPOTHESIS TEST: WHAT IS IT?

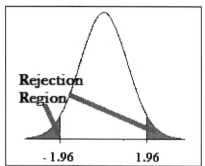

Hypothesis testing is probably the most confusing topic you'll come across in elementary statistics. But it's also one of the most important concepts. Unfortunately, most textbooks don't explain hypothesis tests for non-math majors. You'll find definitions like:

"...method of statistical inference using data from a scientific study" (Wikipedia).

"Hypothesis testing is the use of statistics to determine the probability that a given hypothesis is true." (Wolfram).

"Significance testing is used to help make a judgment about a claim by addressing the question, Can the observed difference be attributed to chance?" (San Jose State University).

The third definition is getting close to explaining what a hypothesis test is in simple terms. The concept actually *is* simple:

*A hypothesis test helps you figure out if you're **right** or **wrong** with results from a test or an experiment.*

For example, let's say you run some figures and figure out your college degree is going to net you an extra $1 million over your lifetime. Reason would tell you that the $1 million figure isn't *exactly* accurate. But how accurate is it? Are you off by a few dollars or are you **way off**? Maybe your friend runs the same test and gets an extra $2 million. Who is right? Who is wrong? Running a hypothesis test on your data will tell you **who is probably right**.

Notice that I said "probably" in the last sentence. That's because nothing in statistics is *exactly* right or *exactly* wrong. Even if you ran a hypothesis test and found out your $1 million figure is valid, you can never be sure that figure is exact. It could be a million dollars and a penny, or $999,900. It's close enough for you to be confident (maybe 99% confident).

Parts of a Hypothesis Test.

A hypothesis test is made up of four stages:

1. **Stating the hypothesis**. This is just saying *what you're testing*. In the salary example, you might state that "I expect to earn a million extra dollars from my degree." **See**: How to State the Null Hypothesis.

2. **Calculate a test statistic**. This step is a little more complex, because there are several different types of test statistics. The good news is that in elementary stats you'll usually be told what test statistic to use. For example, if you have normally distributed data and a fairly large sample you can use a z-score. The test statistic will tell you if your results are valid (i.e. if you're probably right, or probably wrong).

3. **Convert the test statistic from Step 3 into a P-Value.** This gives you a probability that you can use to decide if you're right, or wrong. More good news: Excel and other software can calculate test statistics and p-values for you.

4. **Use the p-value** to decide to support or reject the null hypothesis (your research statement).

P-Values and Alpha Levels

What is a P-Value?

A p-value is used in hypothesis testing to help you support or reject the null hypothesis. The p-value is the evidence **against** a null hypothesis. P-values are expressed as decimals although it may be easier to understand what they are if you convert them to a percentage. For example, a p-value of 0.0254 is 2.54%.

When you run a hypothesis test, you compare the p-value from your test to the alpha level you selected when you ran the test. Alpha levels can also be written as percentages.

P-Values vs Alpha levels

Alpha levels are controlled by the researcher (or you) and are related to confidence intervals. You get an alpha level by subtracting your confidence interval from 100%. For example, if you want to be 98 percent confident in your research, the alpha level would be 2% (100% − 98%). When you run the hypothesis test, the test will give you a p-value. Compare that value to your chosen alpha level. For example, let's say you chose an alpha level of 5% (0.05). If the results from the test give you:

A small p-value (≤ 0.05), reject the null hypothesis. This is strong evidence that the null hypothesis is invalid.

A large p-value (> 0.05) means the alternate hypothesis is weak, so you do not reject the null hypothesis.

P-Values and Critical Values

The p-value is just one piece of information you can use when deciding if your null hypothesis is true or not. You can use other values given by your test to help you decide. For example, if you run an f test two sample for variances in Excel, you'll get a p-value, an f-critical value and a f-value.

F	1.610145			
P(F<=f) or	0.244531		Compare the two f-values	
F Critical c	3.178893			

In the above image, the results from the f-test show a large p-value (.244531, or 24.4531%), so you would not reject the null. However, there's also another way you can decide: compare your f-value with your f-critical value. If the f-critical value is smaller than the f-value, you should reject the null hypothesis. In this particular test, the p-value *and* the f-critical values are both very large so you do not have enough evidence to reject the null hypothesis.

REJECTION REGION FOR STATISTICAL TESTS

What is a Rejection Region?

The main purpose of statistics is to test theories or results from experiments. For example, you might have invented a new fertilizer that you think makes plants grow 50% faster. In order to prove your theory is true, your experiment must:

1. Be repeatable.
2. Be compared to a known fact about plants (in this example, probably the average growth rate of plants *without* the fertilizer).

We call this type of statistical testing a hypothesis test. The rejection region is a part of the testing process. Specifically, the rejection region is an area of probability that tells you if your theory (your "hypothesis") is probably true, or probably not true.

Rejection Regions and P-Values.

When you run a hypothesis test (for example, a z test), the result of that test will be a p value. The p value is a "probability value." It's what tells you if your hypothesis statement is probably true or not. If the p value falls in the rejection region, it means you have statistically significant results; You can reject the null hypothesis. If the p-value falls outside the rejection region, it means your results aren't enough to throw out the null hypothesis. What is **statistically significant**? In the example of the plant fertilizer, a statistically significant result would be one that shows the fertilizer does indeed make plants grow faster (compared to other fertilizers).

Rejection Regions and Probability Distributions

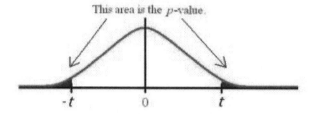

The rejection regions in a two-tailed t-distribution. Image: ETSU.edu

Every rejection region can be drawn on a probability distribution. The image to the left shows a t-distribution with a two-tailed rejection region. It's also possible to have a rejection region in one tail only.

Two Tailed vs. One Tailed Rejection Regions

Which type of test is determined by your null hypothesis statement. For example, if your statement asks *"Is the average growth rate greater than 10cm a day?"* that's a one tailed test, because you are only interested in one direction (greater than 10cm a day). You could also have a single rejection region for "less than". For example, "Is the growth rate less than 10cm a day?" A two tailed test, with two rejection regions, would be used when you want to know if there's a difference in both directions (greater than **and** less than).

Rejection Regions and Alpha Levels

You, as a researcher, choose the alpha level you are willing to accept. For example, if you wanted to be 95% confident that your results are significant, you would choose a 5% alpha level (100% –

95%). That 5% alpha level is the **rejection region**. For a one tailed test, the 5% would be in one tail. For a two tailed test, the rejection region would be in two tails.

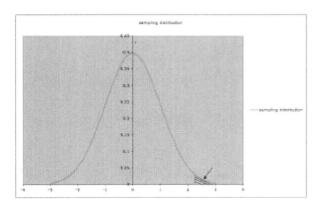

A one tailed test with the rejection region in one tail.

HOW TO FIND A CRITICAL VALUE IN 10 SECONDS (TWO-TAILED TEST)

Let's say you have an alpha level of .05 for this example.

Step 1: *Subtract alpha from 1.*

1 - .05 = **.95**

Step 2: *Divide Step 1 by 2 (because we are looking for a two-tailed test):*

$$.95 / 2 = .475$$

Step 3: *Look at your z-table and locate the alpha level in the middle section of the z-table.* The fastest way to do this is to use an online z-table and use the find function of your browser (usually CTRL+F). In this example we're going to look for .475, so go ahead and press CTRL+F, then type in .475.

Step 4: In this example, you should have found the number .4750. Look to the far left or the row, you'll see the number 1.9 and look to the top of the column, you'll see .06. Add them together to get **1.96**. That's the critical value!

Tip: The critical value appears twice in the z table because you're looking for both a left hand and a right hand tail, so don't forget to add plus or minus! In our example you'd get **±1.96**.

HOW TO DECIDE IF A HYPOTHESIS TEST IS ONE TAILED OR TWO TAILED

In a hypothesis test, you are asked to decide if a claim is true or not. For example, if someone says "all Floridians have a 50% increased chance of melanoma," it's up to you to decide if this claim holds merit. The first step is to look up a z-value, and in order to do *that*, you need to know if it's a one-tailed test or a two-tailed test.

Luckily, you can figure this out in just a couple of steps.

Sample problem 1: "A government official claims that the dropout rate for local schools **is 25%.** Last year, 190 out of 603 students dropped out. Is there enough evidence to reject the government official's claim?"

Sample problem 2: "A government official claims that the dropout rate for local schools **is less than 25%.** Last year, 190 out of 603 students dropped out. Is there enough evidence to reject the government official's claim?"

Sample problem 3: "A government official claims that the dropout rate for local schools **is greater than 25%.** Last year, 190 out of 603 students dropped out. Is there enough evidence to reject the government official's claim?"

Step 1: *Rephrase the claim in the question with an equation:*

Sample problem 1: Dropout rate = 25%
Sample problem 2: Dropout rate < 25%
Sample problem 3: Dropout rate > 25%

Step 2: *If Step 1 has an equal sign in it, this is a two-tailed test. If it has > or <, it's a one-tailed test.*

HOW TO DECIDE IF A HYPOTHESIS TEST IS A LEFT TAILED TEST OR A RIGHT TAILED TEST

In a hypothesis test that you know is one-tailed, you need to decide if it's a left-tailed test or a right-tailed test in order to figure out the z-score.

Sample hypothesis 1: Dropout rate < 25%
Sample hypothesis 2: Dropout rate > 75%

Step 1: *Draw a normal distribution curve.* Assume the area under the curve equals 100%, then shade in the related area. Choose the picture that looks most like the picture you've drawn, and then read the information underneath that picture.

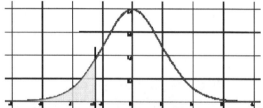

The shaded area could illustrate the area less than 25%. From this diagram you can clearly see that it is a left-tailed test (because the shaded area is on the *left*).

If the shaded area is on the *right*, it's a right-tailed test.

RIGHT TAILED TEST: WHAT IT IS AND HOW TO RUN ONE

The right tailed test is an example of a **one-tailed test**. They are called "one tailed" tests because the rejection region (the area where you would reject the null hypothesis) is only in one tail.

Here's what the right tailed test looks like on a graph:

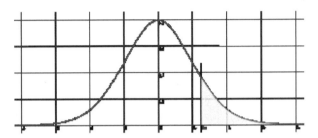

As you can see, the rejection region (shaded) is to the right of the graph.

A right tailed test (sometimes called an upper test) is where your hypothesis statement contains a greater than (>) symbol. In other words, the inequality points to the right. For example, you might be comparing the life of batteries before and after a manufacturing change. If you want to know if the battery life is greater than the original (let's say 90 hours), your hypothesis statements might be:
Null hypothesis: No change ($H_0 = 90$).
Alternate hypothesis: Battery life has increased (H_1) > 90.

The important factor here is that the **alternate hypothesis**(H_1) determines if you have a right tailed test, *not the null hypothesis.*

Right Tailed Test Example.

A high-end computer manufacturer sets the retail cost of their computers based in the manufacturing cost, which iaverages $1800. However, the company thinks there are hidden costs and that the average cost to manufacture the computers is actually much more. The company randomly selects 40 computers from its facilities and finds that the mean cost $\bar{X}$ to produce a computer is $1950 with a standard deviation σ of $500. Run a hypothesis test to see if this thought is true.

Step 1: Write your hypothesis statement (see: the next article, How to state the null hypothesis).
H_0: $\mu \leq 1800$
H_1: $\mu > 1800$

Step 2: Find the test statistic using the z-score formula:

$$z = \frac{\bar{\bar{X}} - \mu}{\sigma/\sqrt{n}}$$

z = 1950 − 1800 / (500/√40) = 1.897

Step 3: Choose an alpha level. No alpha level is mentioned in the question, so use the standard (0.05). Subtract this from 1.
1 − 0.05 = .95
Look up that value (.95) in the middle of the z-table. The area corresponds to a z-value of 1.96. That means you would reject the null hypothesis if your test statistic is greater than 1.96.

1.897 is **not** greater than 1.96, so you cannot reject the null hypothesis.

HOW TO STATE THE NULL HYPOTHESIS – PART ONE

You'll be asked to convert a word problem into a hypothesis statement that will include a null hypothesis and an alternate hypothesis.

Sample Problem: A researcher thinks that if knee surgery patients go to physical therapy twice a week (instead of 3 times), their recovery period will be longer. Average recovery time for knee surgery patients is 8.2 weeks.

Step 1: *Figure out the hypothesis from the problem.* The hypothesis is usually hidden in a word problem, and is sometimes a statement of what you expect to happen in the experiment. The hypothesis in the above question is "I expect the average recovery period to be greater than 8.2 weeks."

Step 2: *Convert the hypothesis to math.* Remember that the average is sometimes written as μ:

$H_1: \mu > 8.2$

That's H_1 (the hypothesis): μ (the average) > (is greater than) 8.2

Step 3: *State what will happen if the hypothesis doesn't come true.* If the recovery time isn't greater than 8.2 weeks, there are only two possibilities, that the recovery time is equal to 8.2 weeks or less than 8.2 weeks.

$H_0: \mu \leq 8.2$

That's H_0 (the null hypothesis): μ (the average) ≤ (is less than or equal to) 8.2

But what if the researcher doesn't have any idea what will happen? See part two of Stating the Null and Alternate Hypothesis.

HOW TO STATE THE NULL HYPOTHESIS (PART TWO)

In the previous section on How to State the Null Hypothesis, I explained how to convert a word problem into a hypothesis statement if you have an idea about what the result will be (i.e. you think you know if the mean will be greater or smaller). But how do you state the null hypothesis if you have no idea about the experiment outcome?

Sample Problem: A researcher is studying the effects of radical exercise program on knee surgery patients. There is a good chance the therapy will improve recovery time, but there's also the possibility it will make it worse. Average recovery time for knee surgery patients is 8.2 weeks.

Step 1: *State what will happen if the experiment doesn't make any difference.* That's the null hypothesis: nothing will happen. In this experiment, if nothing happens, then the recovery time will stay at 8.2 weeks.

$H_0: \mu = 8.2$

That's H_0 (the null hypothesis): μ (the average) = (is equal to) 8.2

Step 2: *Figure out the alternate hypothesis.* The alternate hypothesis is the opposite of the null hypothesis. In other words, what happens if our experiment makes a difference?

$H_1: \mu \neq 8.2$

That's H_1 (The alternate hypothesis): μ (the average) $\neq$ (is not equal to) 8.2

HOW TO SUPPORT OR REJECT A NULL HYPOTHESIS

Your task is to decide whether the evidence supports a claim. If you have a P-value, or are asked to find a P-value, do not follow these instructions; instead go to the section "Null Hypothesis–P-Value."

Step 1: *State the null hypothesis and the alternate hypothesis ("the claim").* If you aren't sure how to do this, review the section "How to State the Null and Alternate Hypothesis."

Step 2: *Find the critical value.* If you don't know how, review the section "How to find a critical value in 10 seconds."

Step 3: *Draw a normal distribution including your critical value.* Shade in the appropriate area (i.e. for a right- or left-tailed test, shade in the tail). This step is optional: it may help you visualize the problem.

Step 4: *If you already have the population standard deviation, go straight to Step 6. If you do not have the population standard deviation (i.e. you are presented with a list of data) go to Step 5.*

Step 5: *Find the sample mean $\bar{X}$ and sample standard deviation (s) for the data.* For information on finding the mean, see: "How to Find the Mean, Mode, and Median." For information on finding the standard deviation, see: "How to Find the Sample Variance and Standard Deviation."

Step 6: *Use the following formula to find the z-value.* If you are given the population standard deviation σ, use it instead of the sample standard deviation, *s.*

$$z = \frac{\bar{X} - \mu}{s/\sqrt{n}}$$

1. Subtract the null hypothesis mean (μ–you figured this out in Step 1) from the mean of the data ($\bar{X}$ from Step 5). Set this number aside for a moment.

2. Divide the standard deviation (σ or s) by the square root of your sample size (the question either stated your sample size or you were given a certain amount n of data). For example, if thirty six children are in your sample and your standard deviation is 2, then:
$3/\sqrt{36} = .5$.

3. Divide your result from Step 1 by your result from Step 2.

Step 7: *Compare your answer from Step 6 with the α value given in the question.* If Step 6 is less than α, reject the null hypothesis, otherwise do not reject it.

HOW TO SUPPORT OR REJECT A NULL HYPOTHESIS (USING A P-VALUE)

If you do not have a P-Value, instead look in the above section "How to Reject or Support a Null Hypothesis."

Step 1: *State the null hypothesis and the alternate hypothesis ("the claim").* If you aren't sure how to do this, see the section "How to State the Null Hypothesis" parts 1 and 2.

Step 2: *Find the critical value. If you don't know how, see the section "How to Find a Critical Value in 10 Seconds."*

Step 3: *Draw a normal distribution with your critical value.* Shade in the appropriate area (i.e. for a right- or left-tailed test, shade in the tail).

Step 4: *If you already have the population standard deviation, go straight to Step 6. If you do not have the population standard deviation (i.e. you are presented with a list of data) go to Step 5.*

Step 5: *Find the mean $(\overline{X})$ and standard deviation (s) for the data.* For information on finding the mean, look in Chapter 1, "How to Find the Mean, Mode, and Median." For information on finding the standard deviation, look in Chapter 2, "How to Find the Sample Variance and Standard Deviation."

Step 6: *Use the following formula to find the z-value. If you have* the population standard deviation σ, use it instead of the sample std dev, *s*.

$$z = \frac{\bar{X} - \mu}{s/\sqrt{n}}$$

Here's how to solve the formula:

1. Subtract the null hypothesis mean (μ–you figured this out in Step 1) from the mean of the sample data $(\bar{X})$ from Step 5).
2. Divide the standard deviation (σ or s) by the square root of your sample size (the question either stated your sample size or you were given a certain amount n of data). For example, if thirty six children are in your sample and your standard deviation is 2, then 3/√36 = .5.
3. Divide your result from Step 1 by your result from Step 2.

Step 7: *Find the P-Value by looking up your answer from Step 6 in the z-table (see resources at the back of this book).* Note: for a two- tailed test, you'll need to double this amount to get the P-Value.

Step 8: *Compare your answer from Step 7 with the α value given in the question.* If Step 7 is less than α, reject the null hypothesis, otherwise do not reject it.

HOW TO SUPPORT OR REJECT A NULL HYPOTHESIS (FOR A PROPORTION)

If you have specific numbers instead of a proportion, see the section "How to Support or Reject a Null Hypothesis," parts 1 and 2.

Sample Problem: A researcher claims that 16% of vegetarians are actually vegans. In a recent survey, 19 out of 100 vegetarians stated they were vegan. Is there enough evidence at α = .05 to support this claim?

Step 1: *State the null hypothesis and the alternate hypothesis ("the claim").*

H_0: p = .16

H_1: p ≠ .16
If you aren't sure how to do this, follow the instructions in the section "How To State the Null Hypothesis."

Step 2: *Find the critical value. If you don't know how, see section "How to Find a Critical Value in 10 seconds."*

Step 3: *Compute $\hat{p}$* by dividing the number of positive respondents from the number in the random sample:

19 / 100 = **.19**

Step 4: *Find p by converting the stated claim to a decimal:*

16% = **.16**
Step 5: Find 'q' by subtracting p from 1:

$$1 - .16 = .84$$

Step 6: *Use the following formula to calculate your test value:*

$$z = \frac{\hat{p} - p}{\sqrt{(p\,q)/n}}$$

1. Subtract p from $\hat{p}$:
.19 - .16 = **.03**
2. Multiply p and q together, then divide by the number in the random sample:
(.16 × .84) / 200 = **.000672**
3. Take the square root of your answer from (2):
√.000672 = **.0259**
4. Divide your answer to (1) by your answer in (3):
.03 / .0259 = **1.158**

Step 6: *Compare your answer from Step 5 with the α value given in the question.* If Step 5 is less than α, reject the null hypothesis, otherwise do not reject it. In this case, 1.158 is not less than our α, so we **do not reject the null hypothesis.**

HOW TO SUPPORT OR REJECT A NULL HYPOTHESIS (FOR A PROPORTION: P VALUE METHOD)

Sample Problem: A researcher claims that more than 23% of community members go to church regularly. In a recent survey, 126 out of 420 people stated they went to church regularly. Is there enough evidence at α = .05 to support this claim? Use the P-Value method.

Step 1: *State the null hypothesis and the alternate hypothesis ("the claim"):*

H_0: p ≤ .23

H_1: p > .23 (claim)

If you aren't sure how to do this, see the section "How To State the Null Hypothesis."

Step 2: *Compute* $\hat{p}$ *by dividing the number of positive respondents from the number in the random sample:*

63 / 210 = **.3**

Step 3: *Find p by converting the stated claim to a decimal:*

23% = **.23**

Step 4: *Find q by subtracting p from 1:*

1 - .23 = **.77**

Step 5: *Use the following formula to calculate your test value:*

$$z = \frac{\hat{p} - p}{\sqrt{(p\,q)/n}}$$

1. Subtract p from $\hat{p}$
.3 - .23 = **.07**
2. Multiply p and q together, then divide by the number in the random sample:
(.23 × .77) / 420 = **.00042**
3. Take the square root of your answer to (2):
√.1771 = **.0205**
4. Divide your answer to (1) by your answer in (3):
.07 / .0205 = **3.41**

Step 6: *Find the P-Value by looking up your answer from Step 5 in the z-table (see resources in the back of this book).* The z-value for 3.41 is .4997. Subtract from .500:

.5 - .4977 = **.023**

Step 7: *Compare your P-value to α.* If the P-value is less, reject the null hypothesis. If the P-value is more, keep the null hypothesis. .023 < .05, so **we have enough evidence to reject the null hypothesis and accept the claim.**

T SCORE IN STATISTICS: EASY CALCULATION STEPS
What is a T Score?

A t score is similar to a z score — it represents **the number of standard deviations from the mean**. T-scores are generally used when you don't know the population standard deviation or the sample size is less than 30. While the z-score returns values from between -5 and 5 (most scores fall between -3 and 3) standard deviations from the mean, the t-score has a greater value and returns results from between 0 to 100 (most scores will fall between 20 and 80). Many people prefer t-scores because the lack of negative numbers means they are easier to work with and there is a larger range so decimals are almost eliminated. This table shows z-scores and their equivalent t-scores.

z-score	t-score
-5	0
-4	10
-3	20
-2	30
-1	40
0	50
1	60
2	70
3	80
4	90
5	100

Calculating a t score is really just a **conversion** from a z score to a t score, much like converting Celsius to Fahrenheit. The formula to convert a z score to a t score is:

T = (Z x 10) + 50.

Sample question: A candidate for a job takes a written test where the average score is 1026 and the standard deviation is 209. The candidate scores 1100. Calculate the t score for this candidate.

Note: If you are given the z-score for a question, skip to Step 2.

Step 1: Calculate the z-score. (See: the z-score formula). The z-score for the data in this sample question is .354.

Step 2: Multiply the z-score from Step 1 by 10:
10 * .354 = 35.4.

Step 3: Add 50 to your result from **Step 2:**
35.4 + 50 = 85.4.

That's it!

Tips:

Z-scores and t-scores both represent standard deviations from the mean, but while "0" on a z-score is 0 standard deviations from the mean, a "50" on a t-score represents the same thing. That's because t-scores use a mean of 50 and z-scores use a mean of 0.

A t-score of over 50 means above average; below 50 means below average. In general, a t-score of above 60 means that the score is in the top one-sixth of the distribution; above 63, the top one-tenth. A t-Score below 40 indicates a lowest one-sixth position; below 37, the bottom one-tenth.

F TEST: WHAT IT IS AND HOW TO RUN ONE
When is an F Test used?

Several different tests are used in statistics, including the Z Test and the Chi-Square test. An F test is run in a very particular situation: when you want to compare two variances. The test compares the ratio of two variances. If the variances are equal, the ratio of the variances will equal 1. For example, if you had two data sets with a sample 1 variance of 10 and a sample 2 variance of 10, the ratio would be 10/10 = 1. You **always** assume that the population variances are equal when running an F Test. In other words, you always assume that the variances are equal to 1. Therefore, your null hypothesis will always be that *the variances are equal*.

Several **assumptions** are made for the test. Your population **must be approximately normally distributed** (i.e. fit the shape of a bell curve) in order to use the test. Plus, the samples must be independent. In addition, you'll want to bear in mind a few important points:

1. The larger variance should always go in the numerator to force the test into a right-tailed test. Right-tailed tests are easier to calculate.
2. For two-tailed tests, divide alpha by 2 before finding the right critical value.
3. If you are given standard deviations, they must be squared to get the variances.
4. If your degrees of freedom aren't listed in the F Table, use the larger critical value. This helps to avoid the possibility of Type I errors.

The F Test and F Tables.

F Table for alpha=.05

df2/df1	1	2	3	4
1	161.4476	199.5000	215.7073	224.5832
2	18.5128	19.0000	19.1643	19.2468
3	10.1280	9.5521	9.2766	9.1172
4	7.7086	6.9443	6.5914	6.3882
5	6.6079	5.7861	5.4095	5.1922
6	5.9874	5.1433	4.7571	4.5337

The F Table is used to find F critical values for your test. It is actually a *collection* of tables. Which one you choose depends on which alpha level (level of significance) you want. When you perform an F Test, you would decide on an alpha level first (say, 0.05) and then look up the numerator degrees of freedom and the denominator degrees of freedom to find the F critical value.

How to run an F test.

F Tests should always be performed using technology, if at all possible. For example, you can run an F Test two sample to compare variances in Excel. While it's possible to calculate an F test by hand, it's a very tedious process that's prone to errors — especially if you're dealing with large data sets.

HOW TO CONDUCT A STATISTICAL F TEST TO COMPARE TWO VARIANCES

One tailed or two tailed F test?

A one-tailed F test would answer the question: Is one variance bigger (or smaller) than the other? In notation, your alternate hypothesis for a one tailed F test would be:

$H_a = \sigma^2 1 > \sigma^2 2$

or

$H_a = \sigma^2 1 < \sigma^2 2$

With a two tailed F test, you just want to know if the variances are not equal to each other. In notation:

$H_a = \sigma^2 1 \neq \sigma^2 2$

Finding an F Statistic: Sample problem

Step 1: *If you are given standard deviations, go to Step 2. If you are given variances to compare, go to Step 3.*

Step 2: *Square both standard deviations to get the variances.* For example, if $\sigma_1 = 9.6$ and $\sigma_2 = 10.9$, then the variances (s_1 and s_2) would be $9.6^2 =$ **92.16** and $10.9^2 =$ **118.81**.

Step 3: *Take the largest variance, and divide it by the smallest variance.* For example, if your two variances were $s_1 = 2.5$ and $s_2 = 9.4$, divide

$9.4 / 2.5 =$ **3.76**

How to run a two tailed F Test.

The differences between running a one tailed or two tailed F test are:

It doesn't matter which sample has the larger variance, so you can put the larger variance from either sample in the numerator.

The alpha level needs to be doubled for two tailed F tests. For example, instead of working at $\alpha = 0.05$, you use $\alpha = 0.10$; Instead of working at $\alpha = 0.01$, you use $\alpha = 0.02$.

Sample Problem: Conduct a two tailed F Test on the following samples:
Sample 1: Variance = 109.63, sample size = 41.
Sample 2: Variance = 65.99, sample size = 21.

Step 1: Write your hypothesis statements:
H_o: No difference in variances.
H_a: Difference in variances.

Step 2: Calculate your F Statistic:
F Statistic = 109.63 / 65.99 = 1.66

Step 3: Choose an alpha level. An alpha level of 0.05 is standard in statistics. Remember this needs to be doubled, so use 0.10.

Step 4: Find the critical F Value using an F Table (see resources). There are several tables, so make sure you look in the alpha = .1 table. The degrees of freedom in the table will be the sample size -1, so sample 1 has 40 degrees of freedom (the numerator) and sample 2 has 20 degrees of freedom (the denominator):
Critical F (40,20) at alpha (0.1) = 1.71

Linear Regression

HOW TO CONSTRUCT A SCATTER PLOT

A scatter plot gives you a visual clue of what is happening with your data. Scatter plots in statistics create the foundation for linear regression, where we take scatter plots and try to create a usable model using functions. There are just three steps to creating a scatter plot.

Sample Problem: Create a scatter plot for the following data:

x	y
3	25
4.1	25
5	30
6	29
6.1	42
6.3	46

Step 1: *Draw a graph. Label the x- and y- axis. Choose a range that includes the maximums and minimums from the given data. In our example, our x-values go from 3 to 6.3, so a range from 3 to 7 would be appropriate.*

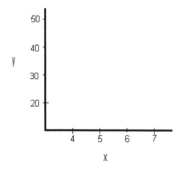

Step 2: *Draw the first point on the graph.* Our first point is (3, 25):

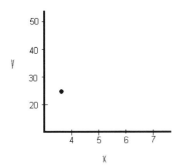

Step 3: *Draw the remaining points on the graph:*

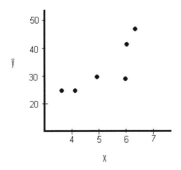

HOW TO COMPUTE PEARSON'S CORRELATION COEFFICIENTS

Correlation coefficients are used in statistics to measure how strong a relationship is between two variables. There are several types of correlation coefficient: Pearson's correlation is a correlation coefficient commonly used in linear regression.

Sample Problem: Compute the value of the correlation coefficient from the following table:

Subject	Age x	Glucose Level y
1	43	99
2	21	65
3	25	79
4	42	75
5	57	87
6	59	81

Step 1: *Make a chart.* Use the given data, and add three more columns: xy, x^2, and y^2:

Subject	Age x	Glucose Level y	xy	x^2	y^2
1	43	99			
2	21	65			
3	25	79			
4	42	75			
5	57	87			
6	59	81			

Step 2.*Multiply x and y to fill the xy column. For example, row 1 would be 43 x 99 = 4257:*

Subject	Age	Glucose	xy	x^2	y^2
1	43	99	4257		
2	21	65	1365		
3	25	79	1975		
4	42	75	3150		
5	57	87	4959		
6	59	81	4779		

Step 3: *Take the square of the numbers in the x column, and put the result in the x^2 column:*

Subject	Age x	Glucose Level y	xy	x^2	y^2
1	43	99	4257	1849	
2	21	65	1365	441	
3	25	79	1975	625	
4	42	75	3150	1764	
5	57	87	4959	3249	
6	59	81	4779	3481	

Step 4: *Take the square of the numbers in the y column, and put the result in the y^2 column:*

Subject	Age x	Glucose Level y	xy	x^2	y^2
1	43	99	4257	1849	9801
2	21	65	1365	441	4225
3	25	79	1975	625	6241
4	42	75	3150	1764	5625
5	57	87	4959	3249	7569
6	59	81	4779	3481	6561

Step 5: *Add up all of the numbers in the columns and put the result at the bottom row.* The Greek letter sigma (Σ) is a short way of saying "sum of" (in other words—add them up!):

Subject	Age x	Glucose Level y	xy	x^2	y^2
1	43	99	4257	1849	9801
2	21	65	1365	441	4225
3	25	79	1975	625	6241
4	42	75	3150	1764	5625
5	57	87	4959	3249	7569
6	59	81	4779	3481	6561
Σ	247	486	20485	11409	40022

Step 6: *Use the following formula to work out the correlation coefficient:*

$$r = \frac{n(\sum xy) - (\sum x)(\sum y)}{\sqrt{[n(\sum x^2) - (\sum x)^2][n(\sum y^2) - (\sum y)^2]}}$$

From our table:

$\sum x$ = 247
$\sum y$ = 486
$\sum xy$ = 20,485
$\sum x^2$ = 11,409
$\sum y^2$ = 40,022
n is the sample size = 6

1. Replace the variables with the above values:

$$r = \frac{6(20,485) - (247)(486)}{\sqrt{[6(11,409) - (247)^2][6(40,022) - (486)^2]}}$$

2. Multiply n by $\sum xy$:

    6 × 20,485 = **122,910**

3. Multiply $\sum x$ by $\sum y$:

    247 × 486 = **120,042**

4. Subtract step (3) from step (2):
    122,910 - 120,042 = **2,868**
5. Multiply n by $\sum x^2$:
    6 × 11,409 = 68,454
6. Square $\sum x$:
    247 × 247 = **61,009**
7. Subtract step (6) from step (5):
    68,454 − 61,009 =7,445
8. Multiply n by $\sum y^2$:
    6 × 40,022 = **240,132**

9. Square Σy:
    486 × 486 = **236,196**
10. Subtract step (9) from (8):
    240,132 − 236,196 = **3,936**
11. Multiply step (7) by step (10):
    7,445 × 3,936 = 29,303,520
11(a). Find the square root of step 11:
    $\sqrt{29,303,520} = 5413.27258$
12. Divide step (4) by step (11):

    2,868 / 5,413.27258 = 0.5298
The range of the correlation coefficient is from -1 to 1, where 1 means that the variables are completely related, -1 means the variables are completely opposite, and 0 means the variables aren't related at all. Our result is **0.5298**, or 52.98%, which means the X and Y variables have a moderate positive correlation.

HOW TO COMPUTE A LINEAR REGRESSION TEST VALUE

Linear regression test values are used in linear regression exactly the same way as they are in hypothesis testing, but instead of working with z-tables you'll be working with t-tables.

Sample Problem: Given a set of data with sample size 8 and r = .454, compute the test value.
Step 1: *Compute r, the correlation coefficient, unless it has already been given to you in the question. In our example, r = .454*
Step 2: *Use the following formula to compute the test value (n is the sample size):*

$$T = r\sqrt{\frac{n-2}{1-r^2}}$$

1. *Replace the variables with your numbers:*

$$T = .454\sqrt{\frac{8-2}{1-.454^2}}$$

2. Subtract 2 from n:
     8 − 2 = **6**
3. Square r:
     .454 × .454 = **.206116**
4. Subtract step (3) from 1:
     1 - .206116 = **.793884**
5. Divide step (2) by step (4):
     6 / .793884 = **7.557779**
6. Take the square root of step (5):
     √7.557779 = **2.74914154**
7. Multiply r by step (6):

   .454 × 2.74914154 = **1.24811026**

HOW TO FIND THE COEFFICIENT OF DETERMINATION

The coefficient of determination is a measure of how well a statistical model is likely to predict future outcomes. The coefficient of determination, r^2, is the square of the sample correlation coefficient between outcomes and predicted values. The correlation coefficient for a sample is .543. What is the coefficient of determination?"

Step 1: *Find the correlation coefficient, r (it may be given to you in the question).* In our example:

r = **.543**

Step 2: *Square the correlation coefficient:*

$.543^2$ = **.295**

Step 3: *Convert the correlation coefficient to a percentage*:

.295 = **29.5%**

HOW TO FIND A LINEAR REGRESSION EQUATION

When a correlation coefficient shows that data is likely to be able to predict future outcomes, statisticians use linear regression to find a predictive function. Recall from elementary algebra, the equation for a line is **y = mx + b**. This article shows you how to take data, calculate linear regression, and find the equation
y' = a + bx.

Step 1: *Make a chart of your data, filling in the columns in the same way as you would fill in the chart if you were finding the Pearson's Correlation Coefficient.* If you don't know how, see "How to Compute Pearson's Correlation Coefficient."

Subject	Age x	Glucose Level y	xy	x^2	y^2
1	43	99	4257	1849	9801
2	21	65	1365	441	4225
3	25	79	1975	625	6241
4	42	75	3150	1764	5625
5	57	87	4959	3249	7569
6	59	81	4779	3481	6561
Σ	247	486	20485	11409	40022

From the above table:

$\Sigma x = 247$
$\Sigma y = 486$
$\Sigma xy = 20485$
$\Sigma x^2 = 11409$
$\Sigma y^2 = 40022$
n is the sample size = 6

Step 2: *Calculate a using the following formula:*

$$a = \frac{(\Sigma y)(\Sigma x^2) - (\Sigma x)(\Sigma xy)}{n(\Sigma x^2) - (\Sigma x)^2}$$

1. Replace the variables with your numbers:

$$a = \frac{(486)(11409) - (247)(20485)}{6(11409) - 247^2}$$

2. Multiply Σy by Σx^2:
   486 × 11,409 = **5,544,774**
3. Multiply Σx by Σxy:
   247 × 20,485 = **5,059,795**
4. Subtract step (3) from step (2):
   5,554,774 − 5,059,795 = **484,979**
5. Multiply n by Σx^2:
   6 × 11,409 = **68,454**
6. Square Σx:
   247 × 247 = **61,009**
7. Subtract step (6) from step (5):

   68,454 − 61,009 = **7,445**

8. Divide step (4) from step (7):
   484,979 / 7,445 = **65.14**

Step 3: *Calculate b using the following formula:*

$$b = \frac{n(\Sigma xy) - (\Sigma x)(\Sigma y)}{n(\Sigma x^2) - (\Sigma x)^2}$$

1. Replace the variable with your numbers:

$$b = \frac{6(20485) - (247)(486)}{6(11409) - 247^2}$$

2. Multiply n by Σxy:
   6 × 20,485 = **122,910**
3. Multiply Σx by Σy:
   247 × 486 = **120,042**
4. Subtract step (3) from step (2):

$122{,}910 - 120{,}042 = \textbf{2{,}868}$

5. Multiply n by Σx^2:

   $6 \times 11{,}409 = \textbf{68{,}454}$

6. Square Σx:

   $247 \times 247 = \textbf{61{,}009}$

7. Subtract step (6) from step (5):

   $68{,}454 - 61{,}009 = \textbf{7{,}445}$

8. Divide step (4) by step (7):

   $2{,}868 \,/\, 7{,}445 = \textbf{.385225}$

Step 3: *Insert the values from steps 2 and 3 into the equation:*

y' = a + bx

y' = 65.14+ .385225x

HOW TO FIND A LINEAR REGRESSION SLOPE

Step 1: *Find the following data from the information given:* Σx, Σy, Σxy, Σx^2, Σy^2. If you don't remember how to get those variables from data, see the section "How to Compute Pearson's Correlation Coefficients." Follow the Steps there to create a table and find Σx, Σy, Σxy, Σx^2, and Σy^2.

Step 2: *Insert the data into the b formula:*

$$b = \frac{n(\Sigma xy) - (\Sigma x)(\Sigma y)}{n(\Sigma x^2) - (\Sigma x)^2}$$

You can find more comprehensive instructions on how to work the formula in the section above, "How to Find a Linear Regression Equation."

Chi Square

WHAT IS A CHI SQUARE TEST?

A chi-square (X^2) test is used to test whether distributions of categorical variables differ from each another.

A **very small chi square test statistic** means that your observed data fits your expected data extremely well.

A **very large chi square test statistic** means that the data does not fit very well. If the chi-square value is large, you reject the null hypothesis.

The Chi-Square Distribution

A chi-square distribution is skewed like a t-distribution. However, while the t-distribution has a minimum value of 0, the chi square distribution has a minimum value of 1.

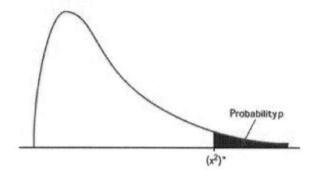

Chi Square P-Values.

A chi square test will give you a p-value. The p-value will tell you if your test results are significant or not. In order to perform a chi square test and get the p-value, you need two pieces of information:

Degrees of freedom. That's just the number of categories minus 1.

The alpha level(α). This is chosen by you, or the researcher. The usual alpha level is 0.05 (5%), but you could also have other levels like 0.01 or 0.10.

In elementary statistics or AP statistics, both the degrees of freedom and the alpha level are usually given to you in a question. You don't normally have to figure out what they are.
You *may* have to figure out the degrees of freedom yourself, but it's pretty simple: count the categories and subtract 1.

Degrees of freedom are placed as a subscript after the chi-square (X^2) symbol. For example, the following chi square shows 6 degrees of freedom:
X^2_6.
And this chi square shows 4 degrees of freedom:
X^2_4.

The chi square test statistic formula.

The formula for the chi-square statistic used in the chi square test is:

$$\chi^2_c = \sum \frac{(O_i - E_i)^2}{E_i}$$

The subscript "c" are the degrees of freedom. "O" is your observed value and E is your expected value. It's very rare that you'll want to actually *use* this formula to find a critical chi-square value by hand! The summation symbol means that you'll have to perform a calculation for every single data item in your data set. As you can probably imagine, the calculations can get very, very, lengthy and tedious. Instead, you'll probably want to use technology, like SPSS or Excel.

HOW TO FIND A CRITICAL CHI SQUARE VALUE FOR A RIGHT TAILED TEST

Chi squared distributions and the associated tables are used in significance testing. This article will tell you in a few short steps how to find a critical chi square value for a right tailed test, given degrees of freedom and the significance level.

Step 1: *Open the chi squared table* (see the resources section at the back of the book). There may be two tables listed: one table is for upper limits and one is for lower limits. You will need to use the upper limit table for the right tailed tests.

Step 2: *Find the degrees of freedom in the left hand column.* The degrees of freedom symbol is the Greek letter v.

Step 3: *Look along the top of the table to find the given significance (i.e. the probability of exceeding the critical value).* The symbol is the Greek letter α.

Step 4: *Look at the intersection of the row in Step 1 and the column in Step 2.* This is the critical chi square value.

Tip: Sometimes you will be given n (the sample size) instead of degrees of freedom. To find degrees of freedom, simply subtract 1 from this number

If n = 6

Then df = 6 – 1 = **5**

HOW TO FIND A CRITICAL CHI SQUARE VALUE FOR A LEFT TAILED TEST

The chi squared statistic is used to compare two sets of data. To find the critical chi square value for a left tailed test, you will be using the table labeled "lower limits."

Step 1: *Subtract your significance level (α) from 1.* For example, if your significance level is .025, then:

1 - .025 = **.975**

Find this value at the top of the chi square table, heading a column.

Step 2: *Find the given degrees of freedom (v) in the left hand column.*

Step 3: *Trace down the significance level column with your finger until you find the row labeled with the given degrees of freedom.* The value you find at this intersection is the critical chi square value for the left-tailed test.

Resources

TABLES

Binomial Distribution Table

http://www.statisticshowto.com/tables/binomial-distribution-table/

Chi Square Table

http://www.statisticshowto.com/tables/chi-squared-distribution-table-right-tail/

F-Table http://www.statisticshowto.com/tables/f-table/

PPMC Critical Values

http://www.statisticshowto.com/tables/ppmc-critical-values/

T-Distribution Table http://www.statisticshowto.com/tables/t-distribution-table/

Z-Table http://www.statisticshowto.com/tables/z-table/

ONLINE CALCULATORS

Binomial Distribution Calculator

http://www.statisticshowto.com/calculators/binomial-distribution- calculator/
Interquartile Range Calculator:
http://www.statisticshowto.com/calculators/interquartile-range-calculator/
Permutation and Combination Calculator

http://www.statisticshowto.com/calculators/permutation-calculator-and-combination-calculator/
Variance and Standard Deviation Calculator

http://www.statisticshowto.com/calculators/variance-and-standard-deviation-calculator/